Lecture Notes in Computer Sci

Commenced Publication in 1973
Founding and Former Series Editors:
Gerhard Goos, Juris Hartmanis, and Jan van Leeuwen

Thierry Declerck Michael Granitzer
Marcin Grzegorzek Massimo Romanelli
Stefan Rüger Michael Sintek (Eds.)

Semantic Multimedia

5th International Conference on Semantic and Digital Media Technologies, SAMT 2010
Saarbrücken, Germany, December 1-3, 2010
Revised Selected Papers

Volume Editors

Thierry Declerck
DFKI GmbH, Language Technology Lab, Saarbrücken, Germany
E-mail: declerck@dfki.de

Michael Granitzer
Know-Center Graz, 8010 Graz, Austria
E-mail: mgrani@know-center.at

Marcin Grzegorzek
University of Siegen, Institute for Vision and Graphics, Siegen, Germany
E-mail: marcin.grzegorzek@uni-siegen.de

Massimo Romanelli
DFKI GmbH, Intelligent User Interfaces Department
Saarbrücken, Germany
E-mail: massimo.romanelli@dfki.de

Stefan Rüger
Knowledge Media Institute, The Open University, Milton Keynes, UK
E-mail: s.rueger@open.ac.uk

Michael Sintek
DFKI GmbH, Knowledge Management Department, Kaiserslautern, Germany
E-mail: michael.sintek@dfki.de

ISSN 0302-9743 e-ISSN 1611-3349
ISBN 978-3-642-23016-5 e-ISBN 978-3-642-23017-2
DOI 10.1007/978-3-642-23017-2
Springer Heidelberg Dordrecht London New York

Library of Congress Control Number: 2011933584

CR Subject Classification (1998): H.5.1, H.4, H.3, I.7, I.4, H.5.2

LNCS Sublibrary: SL 3 – Information Systems and Application, incl. Internet/Web and HCI

Typesetting: Camera-ready by author, data conversion by Scientific Publishing Services, Chennai, India

Printed on acid-free paper

Springer is part of Springer Science+Business Media (www.springer.com)

Preface

We are pleased to welcome you to the proceedings of the 5th International Conference on Semantic and Digital Media Technologies held in Saarbrücken, Germany.

Large amounts of multimedia material, such as images, audio, video, and 3D/4D material, as well as computer-generated 2D, 3D, and 4D content, already exist and are growing at increasing rates. While these amounts are growing, managing, distribution of, and access to multimedia material are becoming ever harder, both for lay and professional users.

The SAMT conference series tackles these problems by investigating the semantics and pragmatics of multimedia generation, management, and user access. The conference targets scientifically valuable research tackling the semantic gap between the low-level signal data representation of multimedia material and the high-level meaning that providers, consumers, and prosumers associate with the content. In a highly selective review procedure only 12 full research papers and 4 short poster papers were selected for inclusion in these proceedings.

At this point, we would like to express our gratitude to all the people who contributed to the technical and organizational success of the SAMT 2010 conference in any kind. Special thanks go to the Publicity Chair, Susanne Thierfelder, who did a great job in the preparation phase for this event. Furthermore, we would like to thank the Program Committee members for a smooth review process, the invited speakers and tutors, as well as all contributors and participants of the conference.

December 2010

Thierry Declerck
Michael Granitzer
Marcin Grzegorzek
Massimo Romanelli
Stefan Rüger
Michael Sintek

Organization

SAMT 2010 Program Committee

Alan Smeaton
Alejandro Jaimes
Alun Evans
Ansgar Scherp
Antoine Isaac
Bernard Mérialdo
Bianca Falcidieno
Brigitte Kerherve
Bruno Bachimont
Carlos Monzo
Cees Snoek
Charlie Cullen
Evan Brown
Chong-Wah Ngo
Christophe Garcia
Craig Lindley
Doug Williams
Ebrahimi Touradj
Ebroul Izquierdo
Elena Simperl
Erik Mannens
Ewald Quak
Frank Nack
Giovanni Tummarello
Harald Kosch
Hyowon Lee
Ichiro Ide
Jacco van Ossenbruggen
Jean Claude Leon
Jeff Z. Pan
Jenny Benois
Susanne Little
Joost Geurts
Jose M. Martinez
Lekha Chaisorn
Li-Qun Xu
Lloyd Rutledge
Marcel Worring
Mark Maybury
Mark Sandler
Mark van Doorn
Mathias Lux
Matthew Cooper
Mauro Barbieri
Michael Granitzer
Michael Hausenblas
Michela Mortara
Michela Spagnuolo
Mohan Kankanhalli
Nadia Thalmann
Nicu Sebe
Nozha Boujemaa
Oscar Celma
Oscar Mayor
Paul Lewis
Paulo Villegas
Philipp Cimiano
Raphael Troncy
Remco Veltkamp
Riccardo Albertoni
Rong Yan
Shin'ichi Satoh
Siegfried Handschuh
Simone Marini
Simone Santini
Stavros Christodoulakis
Stefano Bocconi
Stefanos Kollias
Steffen Staab
Stephane Marchand-Maillet
Susanne Boll
Suzanne Little
Tat-Seng Chua
Thierry Declerck
Thrasyvoulos Pappas
Tobias Buerger
Vasileios Mezaris
Vassilis Tzouvaras
Vincenzo Lombardo
Vojtech Svatek
Willemijn Heeren
William Grosky
Winston Hsu
Wolf-Tilo Balke
Yannis Avrithis
Yiannis Kompatsiaris
Zeljko Obrenovic

Table of Contents

Full Research Papers

Short Poster Papers

Query Expansion in Folksonomies

Rabeeh Abbasi

WeST - Institute for Web Science and Technologies,
University of Koblenz-Landau
Universitätsstrasse 1, 56070 - Koblenz, Germany
abbasi@uni-koblenz.de

Abstract. People share resources in folksonomies and add tags to these resources. There are often only a few tags associated with each resource, which makes the data available in folksonomies extremely sparse. Spareness in folksonomies makes searching resources difficult. Many relevant resources against a query might not be retrieved if they are not associated with the queried terms. One possible way to overcome the search problem in folksonomies is query expansion. We propose and compare different query expansion techniques for folksonomies and evaluate these methods on a large scale. We also propose an automated evaluation method for query expansion. Experimental results show that query expansion helps in improving search in folksonomies.

1 Introduction

Folksonomies allow users to share and tag resources. Tags (keywords) added to the resources provide low-level semantics for the resources. Users can browse and search resources using tags. For example, a user can share images of a Chevrolet car of 1929 model and add the tags *1920s* and *Chevy* to it. The users are free to choose tags. They might not add many relevant tags to the resources. Lack of sufficient number of relevant tags results into sparseness of data and makes the search in Folksonomies difficult. Especially for queries with sparse results. For example if a user is searching for pictures of a 1929 model Chevrolet car using the tags *1929* and *Chevrolet.* He might not be able to retrieve the images tagged only with the tags *1920s* and *Chevy*, although the resources tagged with *1920s* and *Chevy* might also contains images of his interest. We hypothesize that many resources in folksonomies which are unsearchable due to the word mismatch problem, where the queried tags are not associated with the relevant resources.

We can use query expansion techniques to overcome the problem of word mismatch. This paper explores and analyzes different global and local analysis methods for query expansion in folksonomies. We apply existing state-of-the-art techniques like *Pseudo Relevance Feedback* and techniques using external data sources like Wordnet for query expansion. The standard techniques used for query expansion may not perform very well in folksonomies due to data sparseness. Most of the resources in folksonomies are tagged with only a few relevant

T. Declerck et al. (Eds.): SAMT 2010, LNCS 6725, pp. 1–16, 2011.

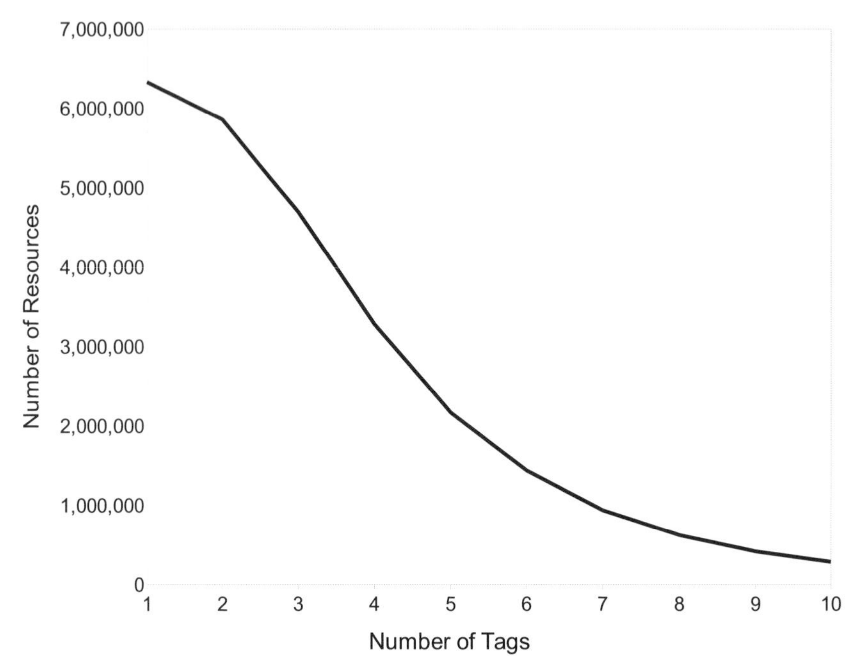

Fig. 1. This figure is based on 27 million images uploaded to Flickr between years 2004 and 2005. Y-axis shows the number of images and x-axis shows the number of tags associated with the images. 6.3 million images are associated with only one tag and only 2.1 million images have five tags.

tags. The spareness in folksonomies can be observed in figure 1 which shows information of 27 million images uploaded to Flickr[1] between years 2004 and 2005. We can see that most of the resources have only a few tags and relatively fewer resources are associated with a higher number of tags. When using a standard query expansion technique like PRF, where top ranked resources are analyzed for query expansion, the top ranked resource might not have sufficient number of tags associated with them to perform query expansion. Therefore we propose and compare different types of query expansion methods.

To evaluate the proposed methods, we propose an automated evaluation method which allows us to evaluate a variety of query expansion techniques and on a large scale. We evaluate 10 different query expansion techniques using automated evaluation on a dataset of 27 million resources and 300 queries. We also propose methods to reduce bias in automated evaluation. Experimental results show that query expansion in folksonomies significantly improves search results, particularly for the queries having only a few relevant resources.

Although research work exists for personalized query expansion [8] and enhancing search in folksonomies using ontologies [25], but to the best of our knowledge, there is no research work focusing on the comparison of different query expansion techniques in folksonomies. The novel contributions of this research work are as follows

[1] flickr.com

- Analyze global and local analysis methods for query expansion in a folksonomy
- Proposes global query expansion techniques based on semantically related tags (Sec. 3.1)
- Exploits Flickr APIs for query expansion (Sec. 3.3)
- Proposes automated evaluation of query expansion

The remainder of this paper is structured as follows, section 2 presents the related work. Section 3 describes the methods used for query expansion in folksonomies. Section 4 contains the description of evaluation setup and dataset used for evaluation. Section 5 describes the results and in section 6 we conclude the presented research work.

2 Related Work

Many researchers have focused on folksonomies or collaborative tagging systems in past the few years. The related research work in folksonomies can be divided into two parts. First, discovering semantics out of tags and second searching and improving resource retrieval in folksonomies.

Extracting Tags Semantics: Query expansion exploits tags relevant to the query for expanding the original query. It is therefore interesting to discuss how researchers analyzed similarities between tags. Discovering semantic relationships between tags has been discussed by [11,15,22,32]. Markines *et al.* [22] provides detailed semantic analysis of tags and verifies the learned semantic relationships using external knowledge base. In tag recommendation systems, (semantically) relevant tags are recommended to the user, in case of query expansion, relevant tags are used for expanding queries. [15,32] propose methods for tag recommendations by finding related tags from within the folksonomy data (by using co-occurrence information etc.). [14,34] use external data sources (e.g. page text, anchor text, Wikipedia etc.) for recommending tags. [23,29] propose methods to extract ontologies from folksonomies.

Searching and Browsing in Folksonomies: Researchers have also focused on improving search and browsing in folksonomies using methods other than query expansion. For example, Jaschke *et al.* [15] presents an algorithm for ranking tags, resources, and users in folksonomies. [1,3,7,18] propose algorithms to improve browsing in folksonomies. They explore semantic and hierarchical relations between tags to improve browsing in folksonomies. Yahia *et al.* [36] presents a study of network-aware search. Zhou *et al.* [38] proposes a generative model for social annotations to improve information retrieval in web search.

Query Expansion: Query expansion is often used to overcome the problem of word mismatch. Related work to query expansion split into two classes: local analysis and global analysis.

In global analysis similar words are identified for expanding the query. Similar words can be obtained from co-occurrence information or from an external data source like Wikipedia, Wordnet or web search results. Wordnet (in addition to other knowledge sources) has been used for finding semantically related words for query expansion by [12,30,33]. Yin *et al.* [37] expands the search query by exploiting top web search result snippets. Collins *et al.* [12] combines multiple data sources (including Wordnet and Co-occurrence data) using a Markov chain framework. Xu *et al.* [35] compares global and local analysis methods for query expansion.

In local analysis for query expansion, similar terms are identified from top relevant documents, which are then used for expanding the original query. Initial research work in local analysis can be found in [5,27,28]. Xu *et al.* [35] utilizes two methods for query expansion using local analysis. In first approach, it expands the query using top relevant documents (local feedback), and in second approach, it selects concepts from the best passages of the top relevant results to expand the query.

In recent years, many researchers have proposed different methods for query expansion. For example, [12,17] propose probabilistic methods for query expansion. [9,13] exploit query logs for expanding queries. [4] proposes different representations of blogs for expanding queries. Bertier *et al.* [8] expands queries using representative tags of a user in folksonomies. Their query expansion method is personalized for the user who gives the query. Pan *et al.* [25] propose to expand folksonomy search using ontologies. Other recent work in query expansion includes [10], which classifies the rare queries using we search results and [31] proposes query expansion for federated search. In our previous work [2], we have focused on enriching the vector space models of a folksonomy by exploiting semantically related tags. In comparison to the work to the research done by [16] about query expansion based on tag co-occurrence, we provide a detailed comparison of different query expansion techniques in folksonomies.

3 Query Expansion

Data in folksonomies is very sparse as compared to the normal text documents, because the number of tags associated to a resource are much less than the words associated to a document. A user searching against a query might not find relevant resources because many relevant resources might be tagged with other semantically related tags. For example, if a user searches for the query "*1940, penny*", he might not find resources tagged with "*forties, penny*" or "*1944, penny*". We can solve this problem of word mismatch by expanding the query with semantically related tags. The commonly used methods for finding semantically related tags use either co-occurrence of tags or external data sources like Wikipedia or Wordnet.

We explore three different approaches for query expansion. First, the co-occurrence based expansion exploits different similarity distributions between tags. Second, using external data sources: Wordnet and Flickr API. Third we expand queries using *Pseudo Relevance Feedback* which is also state of the art in query expansion. Each of these methods is described in detail in the following sub-sections.

3.1 Global Analysis

The basic idea in global analysis for query expansion is that the query term is expanded with relevant terms. The expansion is independent of the queries, that means, a term is expanded with the same relevant terms for different queries. We compare different global analysis methods based on similarity between tags. The relevant tags are identified by measuring their similarity. We classify the similarities between tags into the following three categories:

Co-occurrence: The simplest form of finding similar tags is co-occurrence, two tags are relevant, if they co-occur in resources. Simple co-occurrence between to tags $\mathbf{t}_1$ and $\mathbf{t}_2$ can be computed using Eq. 1, i.e. by counting, in how many resources a tag $\mathbf{t}_1$ appears together with the tag $\mathbf{t}_2$. The main disadvantage of using simple co-occurrence measure is the lack of normalization. Very frequent tags like *sky*, *family* or *travel* co-occur with most of the tags, and it leads to a low quality query expansion. Normalizing co-occurrence with the frequencies of the tags, helps in finding relevant tags which might not be very frequent globally. One way of normalization is using *cosine* similarity (Eq. 2), in which co-occurrence of tags is normalized by the *Euclidean norm* of the tag vectors. We also consider *Dice* (Eq. 3) and *Jaccard* (Eq. 4) co-efficients for finding relevant tags. The *Dice* co-efficient gives higher value to co-occurring tags than the *Jaccard* co-efficient. Jaccard co-efficient penalizes tags which do not co-occur very frequently [20].

$$simple(\mathbf{t}_1, \mathbf{t}_2) = |\mathbf{t}_1 \cap \mathbf{t}_2| \tag{1}$$

$$cosine(\mathbf{t}_1, \mathbf{t}_2) = \frac{\mathbf{t}_1 \cdot \mathbf{t}_2}{\|\mathbf{t}_1\| \cdot \|\mathbf{t}_2\|} \tag{2}$$

$$dice(\mathbf{t}_1, \mathbf{t}_2) = 2\frac{|\mathbf{t}_1 \cap \mathbf{t}_2|}{|\mathbf{t}_1| + |\mathbf{t}_2|} \tag{3}$$

$$jaccard(\mathbf{t}_1, \mathbf{t}_2) = \frac{|\mathbf{t}_1 \cap \mathbf{t}_2|}{|\mathbf{t}_1 \cup \mathbf{t}_2|} \tag{4}$$

Probabilistic: We also use a probabilistic model for finding relevant tags. *Mutual Information* (MI, Eq. 5) measures the association between two tags. The measure of association depends upon the probability of two tags appearing together. If there is a high probability, that the tags $\mathbf{t}_1$ and $\mathbf{t}_2$ appear together, then the value of mutual information between these tags would be high.

$$MI(\mathbf{t}_1, \mathbf{t}_2) = \sum_{\mathbf{t}_1} \sum_{\mathbf{t}_2} p(\mathbf{t}_1, \mathbf{t}_2) \log \left(\frac{p(\mathbf{t}_1, \mathbf{t}_2)}{p(\mathbf{t}_1), p(\mathbf{t}_2)} \right) \tag{5}$$

where $p(\mathbf{t}_1)$ and $p(\mathbf{t}_2)$ are the marginal probabilities of the tags $\mathbf{t}_1$ and $\mathbf{t}_2$ respectively. $p(\mathbf{t}_1, \mathbf{t}_2)$ is the joint probability between the tags $\mathbf{t}_1$ and $\mathbf{t}_2$.

Heuristic: In addition to standard similarity measures (Eq. 1 – 4), we also analyzed two heuristics (*overlap* and *modified overlap coefficients*) for finding relevant tags. The value of the overlap coefficient between two tags $\mathbf{t}_1$ and $\mathbf{t}_2$ is high, if they one the tags mostly appears with the other tag. For example, if a tag *sky* always appear with the tag *blue*, then the value their overlap coefficient is one. The *modified overlap* co-efficient (Eq. 7) is a non-symmetric co-efficient. The value of this coefficient is one for $\mathbf{t}_1$, if $\mathbf{t}_1$ always appear with $\mathbf{t}_2$, but if $\mathbf{t}_2$ does not appear often with $\mathbf{t}_1$, then the modified overlap between $\mathbf{t}_2$ and $\mathbf{t}_1$ would be low. For example, if the tag *sky* always appear with the tag *blue*, then the value of $overlap\_mod(sky, blue)$ would be 1 and if the tag *blue* also appears with many other tags (like *color* or *sea*), then the value of $overlap\_mod(blue, sky)$ would be less than one.

$$overlap(\mathbf{t}_1, \mathbf{t}_2) = \frac{|\mathbf{t}_1 \cap \mathbf{t}_2|}{min(|\mathbf{t}_1|, |\mathbf{t}_2|)} \tag{6}$$

$$overlap\_mod(\mathbf{t}_1, \mathbf{t}_2) = \frac{|\mathbf{t}_1 \cap \mathbf{t}_2|}{|\mathbf{t}_1|} \tag{7}$$

Expansion: We can measure the relevance between tags off-line and expand the original query using pre-computed relevance values. Let us represent the pre-computed similarity matrix with $S = [s_{ij}]_{n \times n}$, where n is the number of tags in the folksonomy and s_{ij} is similarity value between the tags $\mathbf{t}_i$ and $\mathbf{t}_j$ computed using one of the equations from Eq. 1 to Eq. 7. Now we can use the following formula for expanding the original query.

$$\mathbf{q} = S \times \mathbf{q}_0 \tag{8}$$

where $\mathbf{q}$ is the expanded query and $\mathbf{q}_0$ is the original query. The dimension of the queries $\mathbf{q}$ and $\mathbf{q}_0$ is $n \times 1$.

3.2 Local Analysis

The state-of-the-art in automated query expansion is Pseudo Relevance Feedback (PRF) [28,6]. PRF is similar to *Relevance Feedback*, but instead of manually marking the resources as relevant, top documents are considered as relevant to the query and their terms are used for query expansion. Rocchio [27] proposed a method for automated query expansion, in which query features are move towards the centroid of the relevant documents, and away for irrelevant documents. As we do not have irrelevant documents in our case, therefore we expand the query using top relevant documents. The query expansion we used in our experiments is given in the following equation

$$\mathbf{q} = \alpha \mathbf{q}_0 + \beta (\sum R_d) \cdot idf_{R_d} \tag{9}$$

where R_d is the set of relevant documents for the original query $\mathbf{q}_0$ and $\mathbf{q}$ is the expanded query. We limit the set of top relevant documents R_d to 30. idf_{R_d} is the *inverse document frequency* of the tags in the set of relevant documents. We also limited the number of terms used in our experiments to 100 and used *inverse document frequency* to penalize very common tags. We represent this method by *PRF* in the experimental results (Sec. 5).

3.3 External Data Sources

Many researchers has shown that the query expansion using external data sources like Wordnet helps in improving search results [12,30,33,37]. For query expansion in folksonomies, we consider two external data sources for query expansion. First using Wordnet and second using Flickr API.

Table 1. Sample expanded queries using different methods

Method	1929, chevy	lake, huntington	purple, flowers
Co-Occ.	1929, chevy, flood, iowa, cedarrapids, chevrolet, car, cars, automobile	lake, huntington, water, 2005, nature, sunset, beach, garden, california	purple, flowers, flower, green, nature, garden, macro
Cosine	1929, chevy, cedarrapids, christman, memorium, royalyork, chevrolet, chrystler, supercharged, impala	lake, huntington, water, steelhead, frankfort, frenzy, huntingtongardens, hage, skulptur, jardin	purple, flowers, flower, pink, violet, garden, plants, nature
Dice	1929, chevy, cedarrapids, christman, memorium, modela, chevrolet, corvette, impala, carshow	lake, huntington, water, michigan, mountains, tahoe, huntingtongardens, hage, skulptur, jardin	purple, flowers, flower, pink, green, garden, nature, plants
Jaccard	1929, chevy, cedarrapids, christman, memorium, modela, chevrolet, corvette, impala, viper	lake, huntington, water, michigan, fishing, mountains, huntingtongardens, hage, skulptur, jardin	purple, flowers, violet, catchycolors, iris, orchid, garden, flower, plants, macro
MI	1929, chevy, flood, iowa, ford, memorium, chevrolet, car, cars, automobile	lake, huntington, water, tahoe, district, michigan, garden, gardens, beach, giardino	purple, flowers, flower, green, pink, garden, nature, plants
Overlap	1929, chevy, christman, memorium, royalyork, car, chrystler, hhr, supercharged, chevrolet	lake, huntington, peyto, jarvi, natsuki, huntingtongardens, hage, dieu, myths	purple, flowers, roxo, paars, purplerain, beautyberry, daises, genitalia, loreak, gerberas
PRF	chevy, car, 1929, chevrolet, cameraphone, 2004, me, cars, auto, newyork	lake, huntington, sunset, water, 2005, bali, holiday, autumn, nature, fall	flowers, purple, white, yellow, garden, plants, nativeplants, hiking, 2003, macro
Flickr	1929, chevy, chavorlet, car, classic, truck	lake, huntington, water, trees, nature, reflection, beach, california, pier, surf	purple, flowers, flower, pink, blue, nature, spring
Wordnet	1929, chevy	lake, huntington	purple, flowers, flower, blossom, bloom, peak, prime

Wordnet: We use the synsets[24] of Wordnet for query expansion. For any given query tag, we find its synonyms using Wordnet synsets. We only consider the noun synonyms for query expansion. If there is no synonym for a given tag, we do not expand it.

Flickr API: Flickr provides an extensive set of services using APIs. We use the *flickr.tags.getRelated.html*[2] API for expanding the query terms. According to Flickr API website, *flickr.tags.getRelated.html* "returns a list of tags 'related' to the given tag, based on clustered usage analysis". For any given query tag, we expand it using the tags retrieved from the Flickr API.

---

[2] flickr.com/services/api/flickr.tags.getRelated.html

We expanded the query using external data sources using the following equation.

$$\mathbf{q} = \alpha \mathbf{q}_0 + \beta \mathbf{e} \tag{10}$$

where $\mathbf{e}$ is the relevant terms to the queried terms, obtained from external data sources (Wordnet or Flickr). We set $\alpha = 1$ and $\beta = \{0.01, 0.1, 0.5\}$, in our experiments $\beta = 0.1$ has produced the best results for external data sources.

Examples of query expansion using presented methods are shown in table 1 for three different types of queries. The query "*1929, chevy*" has many relevant resources[3], the query "*lake, huntington*" has only a few relevant resources[4] and the common query "*purple, flowers*" has plenty of relevant resources[5]. Due to space limitations, we have listed the top 5 relevant terms for queried term. We can observe that different types of similarities result into different query expansions. The example of Wordnet shows its limitation of limited vocabulary, there are no relevant words for the tag *chevy* in Wordnet, but based on global or local analysis, or an external data source like Flickr API, we can minimize this limitation. We can see that all other methods than Wordnet have the relevant word *Chevrolet* for the tag *chevy*. Next section describes the dataset and evaluation for the proposed query expansion methods.

4 Dataset and Evaluation

In this section we describe the dataset we used for our experiments followed by the evaluation method.

4.1 Data Set

Our large-scale dataset[6] was obtained by systematically crawling the Flickr system during 2006 and 2007. The target of the crawling activity was the core elements, namely users, tags, resources and tag assignments. The statistics of the crawled dataset are summarized in table 2.

Table 2. Flickr dataset statistics

users	tags	resources	tag assignments
319,686	1,607,879	28,153,045	112,900,000

We applied the following strategy to crawl the Flickr data-set. First, we started a tag centric crawl of all photos that were uploaded between January 2004 and December 2005 and that were still present in Flickr as of June 2007. For this

[3] flickr.com/search/?q=1929%20chevy&m=tags&d=posted-20040101-20051231
[4] flickr.com/search/?q=lake%20huntington&m=tags&d=posted-20040101-20051231
[5] flickr.com/search/?q=purple%20flowers&m=tags&d=posted-20040101-20051231
[6] The reference data set used for this evaluation is available at `http://west.uni-koblenz.de/Research/DataSets/PINTSExperimentsDataSets/`

Table 3. Flickr filtered dataset statistics

users	tags	resources	tag assignments
317,260	92,460	26,801,921	94,499,112

purpose, we initialized a list of known tags with the tag assignments of a random set of photos uploaded in 2004 and 2005. After that, for every known tag we started crawling all photos uploaded between January 2004 and December 2005 and further updated the list of known tags. We stopped the process after we reached the end of the list.

We filtered our dataset by removing those tags which were used by less than 10 users. Those users and resources were also removed from the dataset which did not use any tag. In the final dataset, we had data of about 27 million photos, 0.3 million users, and 92 thousand tags. The exact statistics of the dataset are shown in table 3. We did all our experiments on this dataset.

4.2 Automated Evaluation

User based evaluations for large scale experiments have certain disadvantages, e.g., they are expensive (many human hours required), not precisely reproducible, and are not scalable due to limited availability of human resources. We propose a method for automatically evaluating the results of query expansion based on leave-one-out method. Variants of leave-one-out methods have been exploited in many folksonomy related automated evaluations. For example, Jaschke *et al.* [15] and [21] exploit leave-one-out method for evaluating the performance of different tag recommendation systems.

We hypothesize that many relevant resources are not retrieved because of the word mismatch problem, i.e. missing relevant resources are not associated with the queried tags. The missing relevant resources might be tagged with semantically related tags. If the query is correctly expanded, it might contain the semantically related tags which are present in the missing relevant resources, this would result into retrieval of missing relevant resources. We simulate this scenario by randomly removing one of the queried tags from all the resources in the test dataset, then we expand the query, and measure the retrieval performance by comparing the ranked lists of retrieved modified resources and the original test resources. However, the results can be affected by certain bias in training the expansion model. For example, if the training and test resources are tagged by the same user, it is possible that the query expansion model learns the similarity between tags specific to the user and hence produce biased results for that user. To overcome this bias, we split the training and test datasets based on users, that is, resources tagged by a particular user are either in training dataset or test dataset. No resource in test dataset is tagged by the user who has also tagged a resource in the training dataset. We used tagging data from 80% of the users as training dataset and rest as test dataset. Another bias could occur due to the selection of queries from the data itself, to overcome this bias, we selected the queries from AOL query log[7] [26]. AOL query log originally contained around 20 Million queries from 650K users during three months from March to May 2006. Out of these 20M queries, we selected queries for which the user had clicked on a link to the Flickr website. We removed stop words from the queries and also the tags which are specific to Flickr like (pictures, photos, pics, images, etc.). We randomly selected 300 queries, split into three sets of 100 queries each having 1 to 10, 11 to 50 and more than 50 exact matches in the test data set. The average length of the queries was 2.1 tags.

4.3 Evaluation Measures

We evaluate the proposed query expansion methods using three measures (*Precision at k*, *R-Precision*, and *Mean Average Precision*). Precision at k is the precision of relevant resources ranked in the top k search results. It is calculated as follows

$$\mathbf{P@k} = \frac{\#\ \textit{relevant resource in top k results}}{k} \tag{11}$$

R-Precision is similar to $P@k$, but instead of evaluating top k search results, top r search results are evaluated, where r is equal to the number of relevant resources for the query. This enables us to evaluate both the precision and recall (at level r). The value of r can be different for each query. R-Precision is calculated as follows

$$\mathbf{R\text{-}Prec} = \frac{\#\ \textit{relevant resources in to top r results}}{r} \tag{12}$$

In addition to *R-Precision*, we also measure the standard *Mean Average Precision* (MAP) for evaluation. MAP is the mean of all average precisions of individual queries. Average precision is the average of the precision value obtained for top k documents existing after each relevant document is retrieved [19]. If $q \in Q$ is the set of queries and R_{r_q} is the set of relevant documents for the query q, then MAP can be calculated as follows

$$\mathbf{MAP} = \frac{1}{|Q|} \sum_{q=1}^{|Q|} \frac{1}{|R_{r_q}|} \sum_{r=1}^{|R_{r_q}|} Precision(R_{r_q}) \tag{13}$$

In next section we discuss the results.

5 Results and Discussion

Automated evaluation (Sec. 4.2) allows us to do experiments on a large scale, with plenty of different methods for query expansion. We analyze all the query expansion methods described in Section 3 on 300 randomly selected queries. The queries are divided three sets of 100 queries each. Each set having 1 to

[7] We did not use or analyze any personal information from the query log.

10, 11 to 50, or more than 50 (fuzzy) relevant resources (having all the queried tags). As described in Section 4.2, for each query, we randomly remove one of the queried tags from all the test resources and retrieve resources after query expansion. This simulates the word mismatch problem. The original relevant resources (having all the queried tags) would be retrieved, if the expanded query contains the tags associated with the relevant resources. As we remove a queried tag from the test resources, it is not possible to computed the exact precision and recall values. Instead, the results presented for automated evaluation give a comparative overview of different methods. We use cosine similarity between resources and queries for ranking search results. In case of no query expansion (*No Exp.*), only those resources are retrieved which contain at least one of the queried tags.

Table 4 presents the results of queries having 1 to 10 relevant resources. The best results for each type of query are in bold. The baseline method PRF expands queries using top relevant documents, and for queries with only a few relevant resources, the query expansion using PRF remains worst than no query expansion. On the other side, global analysis methods *cosine* (Eq. 2) and *dice* (Eq. 3) produce the best results with 58% improvement over the baseline method (*Pseudo Relevance Feedback*, PRF) and 29% improvement over no expansion (*No Exp.*). Improper query expansion, like in case of simple co-occurrence (Eq. 1, *Co-Occ.*) leads to poor search results. As the query is expanded without any normalization, many irrelevant results are ranked higher in the search results, thus leading to very low performance. Probabilistic query expansion using *Mutual Information* (Eq. 5, *MI*) does not perform well in folksonomies. Due to very sparse data, it becomes difficult to infer many useful relationships using probabilistic methods. Heuristics like *overlap coefficient* (Eq. 6) do not show significant improvement for queries with sparse results. Query expansion using external data source depends upon the type of data source. When using Wordnet, the results are similar to the results without query expansion. Wordnet has a limited vocabulary, and most of the queries remain un-expanded. Flickr API for query expansion gives results similar to that of *cosine* or *dice*.

Results of queries having 11 to 50 relevant resources are shown in table 5. The baseline method PRF gives the best results, because there are sufficient top ranked resources to learn good query expansion. Only simple co-occurrence and mutual information degrade the results when comparing results of all the methods with the results of no query expansion (*No Exp.*). The reason for MI being the spareness in the data to infer meaningful relationships from probabilities. The results for Wordnet are slightly better than no query expansion.

Table 6 presents the results of queries having more than 50 relevant resources. In this case overlap based methods produce best results, with an improvement of 28% and 32% over the baseline method when comparing MAP. When observing top results (P@20), PRF performs significantly better than no query expansion (*No Exp.*), whereas co-occurrence based methods (Eqs. 1 – 4) except *Jaccard*, do not improve results for P@20. The results suggest that Jaccard based query expansion ranks relevant resources higher than other co-occurrence based methods for

Table 4. Results of queries having 1 to 10 relevant resources

	R-PREC	MAP	P@20
No Exp.	0.0368	0.0688	0.0170
PRF	0.0300	0.0475	0.0180
Co-Occ.	0.0112	0.0213	0.0070
Cosine	**0.0474**	**0.0830**	**0.0225**
Dice	**0.0474**	**0.0808**	**0.0225**
Jaccard	0.0294	0.0600	0.0170
MI	0.0112	0.0232	0.0065
Overlap	0.0414	0.0579	0.0185
OverMod	0.0254	0.0401	0.0150
Flickr	0.0441	0.0743	0.0235
Wordnet	0.0368	0.0688	0.0170

Table 5. Results of queries having 11 to 50 relevant resources

	R-PREC	MAP	P@20
No Exp.	0.0640	0.0707	0.0525
PRF	**0.0936**	**0.0944**	**0.0835**
Co-Occ.	0.0163	0.0264	0.0190
Cosine	0.0690	0.0782	0.0600
Dice	0.0680	0.0772	0.0595
Jaccard	0.0836	0.0909	0.0745
MI	0.0256	0.0365	0.0225
Overlap	0.0695	0.0802	0.0635
OverMod	0.0664	0.0757	0.0550
Flickr	0.0677	0.0761	0.0580
Wordnet	0.0644	0.0709	0.0535

Table 6. Results of queries having more than 50 relevant resources

	R-PREC	MAP	P@20
No Exp.	0.0888	0.0986	0.1065
PRF	0.0977	0.0933	0.1540
Co-Occ.	0.0566	0.0534	0.0700
Cosine	0.0965	0.1065	0.1020
Dice	0.0893	0.1017	0.1020
Jaccard	0.1069	0.1121	0.1275
MI	0.0618	0.0659	0.0805
Overlap	**0.1159**	**0.1195**	**0.1850**
OverMod	**0.1215**	**0.1228**	**0.1800**
Flickr	0.0893	0.0998	0.1060
Wordnet	0.0887	0.0985	0.1065

Table 7. Results of queries all the 300 queries

	R-PREC	MAP	P@20
No Exp.	0.0632	0.0794	0.0587
PRF	0.0738	0.0784	0.0852
Co-Occ.	0.0280	0.0337	0.0320
Cosine	0.0710	0.0892	0.0615
Dice	0.0682	0.0866	0.0613
Jaccard	0.0733	0.0877	0.0730
MI	0.0328	0.0419	0.0365
Overlap	**0.0756**	**0.0859**	**0.0890**
OverMod	0.0711	0.0795	0.0833
Flickr	0.0670	0.0834	0.0625
Wordnet	0.0633	0.0794	0.0590

common queries. The overlap coefficients perform better than all other methods for common queries.

Results of all the 300 queries together are presented in table 7. The best results are of *overlap* coefficient, with an improvement of 10% over the baseline. However, these is due to the reason that overlap coefficient performs better for common queries having more than 50 relevant resources and values of the evaluation measures for common queries are relatively larger than other types of queries. When observing results of all the queries, only simple co-occurrence and mutual information based methods perform worst than without query expansion.

Instead of using a single method for query expansion, we propose to use appropriate method depending upon the type of the query. For example, for queries having sparse results we can *cosine* or *dice* based query expansion, for queries having many relevant resources (11 to 50), we can use local analysis methods like *Pseudo Random Feedback*, and for very common queries, we can expand queries using *overlap* coefficient method.

6 Conclusions

In this paper we show that search results can be improved using query expansion. We presented three approaches for query expansion in folksonomies: global analysis, local analysis, and using external data sources. In global analysis, we presented described methods that exploit co-occurrence information, we also presented probabilistic and heuristic methods for query expansion. For local analysis, we used *Pseudo Relevance Feedback* (PRF) which is also state-of-art in automated query expansion. We also used external data sources, Wordnet and Flickr API[8] for query expansion. We proposed a procedure for automated evaluation of query expansion based on *leave-one-out* method. We proposed to minimize the bias of automated evaluation by (1) splitting the data into training and test dataset based on users and (2) selecting queries from a query log, instead of generating them from the data itself. We performed detailed experiments on a large dataset of around 27 Million resources and 300 queries. We used *R-Precision*, *Mean Average Precision*, and *Precision at k* for evaluation. Results show that query expansion improves the search results, especially for queries having only a few relevant resources in the dataset. Based on experimental results, we propose to use *cosine* based expansion for queries with sparse search results, *PRF* for queries having many relevant resources (11 to 50 in experiments), and *overlap* based global expansion for queries having a lot of relevant resources (more than 50 in experiments). In future we plan to develop an online system which combines PRF and global analysis methods for query expansion.

Acknowledgments. I would like to acknowledge Higher Education Commission of Pakistan (HEC) and German Academic Exchange Service (DAAD) for providing scholarship and support for conducting my PhD. I would also like to acknowledge Prof. Dr. Steffen Staab for providing thoughtful comments and ideas related to this research work.

References

1. Abbasi, R., Chernov, S., Nejdl, W., Paiu, R., Staab, S.: Exploiting Flickr Tags and Groups for Finding Landmark Photos. In: Boughanem, M., Berrut, C., Mothe, J., Soule-Dupuy, C. (eds.) ECIR 2009. LNCS, vol. 5478, pp. 654–661. Springer, Heidelberg (2009)
2. Abbasi, R., Staab, S.: RichVSM: enRiched vector space models for folksonomies. In: HT 2009: Proceedings of the 20th ACM Conference on Hypertext and Hypermedia, pp. 219–228. ACM, New York (2009)
3. Abbasi, R., Staab, S., Cimiano, P.: Organizing Resources on Tagging Systems using T-ORG. In: Proceedings of Workshop on Bridging the Gap between Semantic Web and Web 2.0 at ESWC 2007, Innsbruck, Austria, pp. 97–110 (June 2007)
4. Arguello, J., Elsas, J., Callan, J., Carbonell, J.: Document representation and query expansion models for blog recommendation. In: Proc. of the 2nd Intl. Conf. on Weblogs and Social Media (ICWSM). AAAI, Menlo Park (2008)

[8] flickr.com/services/api/flickr.tags.getRelated.html

5. Attar, R., Fraenkel, A.S.: Local feedback in full-text retrieval systems. J. ACM 24(3), 397–417 (1977)
6. Baeza-Yates, R., Ribeiro-Neto, B.: Modern Information Retrieval. Addison Wesley, Reading (1999)
7. Begelman, G., Keller, P., Smadja, F.: Automated Tag Clustering: Improving search and exploration in the tag space. In: Proceedings of the Collaborative Web Tagging Workshop at WWW (2006)
8. Bertier, M., Guerraoui, R., Leroy, V., Kermarrec, A.-M.: Toward personalized query expansion. In: SNS 2009: Proceedings of the Second ACM EuroSys Workshop on Social Network Systems, pp. 7–12. ACM, New York (2009)
9. Billerbeck, B., Scholer, F., Williams, H.E., Zobel, J.: Query expansion using associated queries. In: CIKM 2003: Proceedings of the Twelfth International Conference on Information and Knowledge Management, pp. 2–9. ACM, New York (2003)
10. Broder, A.Z., Fontoura, M., Gabrilovich, E., Joshi, A., Josifovski, V., Zhang, T.: Robust classification of rare queries using web knowledge. In: SIGIR 2007: Proceedings of the 30th Annual International ACM SIGIR Conference on Research and Development in Information Retrieval, pp. 231–238. ACM, New York (2007)
11. Cattuto, C., Benz, D., Hotho, A., Stumme, G.: Semantic Grounding of Tag Relatedness in Social Bookmarking Systems, pp. 615–631 (2008)
12. Collins-Thompson, K., Callan, J.: Query expansion using random walk models. In: CIKM 2005: Proceedings of the 14th ACM International Conference on Information and Knowledge Management, pp. 704–711. ACM, New York (2005)
13. Cui, H., Wen, J.-R., Nie, J.-Y., Ma, W.-Y.: Probabilistic query expansion using query logs. In: WWW 2002: Proceedings of the 11th International Conference on World Wide Web, pp. 325–332. ACM, New York (2002)
14. Heymann, P., Ramage, D., Garcia-Molina, H.: Social tag prediction. In: SIGIR 2008: Proceedings of the 31st Annual International ACM SIGIR Conference on Research and Development in Information Retrieval, pp. 531–538. ACM, New York (2008)
15. Jaschke, R., Marinho, L., Hotho, A., Schmidt-Thieme, L., Stumme, G.: Tag Recommendations in Folksonomies. In: Kok, J.N., Koronacki, J., Lopez de Mantaras, R., Matwin, S., Mladenič, D., Skowron, A. (eds.) PKDD 2007. LNCS (LNAI), vol. 4702, pp. 506–514. Springer, Heidelberg (2007)
16. Jin, S., Lin, H., Su, S.: Query expansion based on folksonomy tag co-occurrence analysis. In: 2009 IEEE International Conference on Granular Computing, pp. 300–305. IEEE, Los Alamitos (2009)
17. Lafferty, J., Zhai, C.: Document language models, query models, and risk minimization for information retrieval. In: SIGIR 2001: Proceedings of the 24th Annual International ACM SIGIR Conference on Research and Development in Information Retrieval, pp. 111–119. ACM, New York (2001)
18. Li, R., Bao, S., Yu, Y., Fei, B., Su, Z.: Towards effective browsing of large scale social annotations. In: Proceedings of the 16th International Conference on World Wide Web, pp. 943–952. ACM Press, New York (2007)
19. Manning, C.D., Raghavan, P., Schütze, H.: Introduction to Information Retrieval. Cambridge University Press, Cambridge (2008)
20. Manning, C.D., Schütze, H.: Foundations of Statistical Natural Language Processing. The MIT Press, Cambridge (1999)
21. Marinho, L.B., Schmidt-Thieme, L.: Collaborative tag recommendations. In: Data Analysis, Machine Learning and Applications. Studies in Classification, Data Analysis, and Knowledge Organization, pp. 533–540. Springer, Heidelberg (2008)

22. Markines, B., Cattuto, C., Menczer, F., Benz, D., Hotho, A., Stumme, G.: Evaluating similarity measures for emergent semantics of social tagging. In: WWW 2009: Proceedings of the 18th International Conference on World Wide Web, pp. 641–650. ACM, New York (2009)
23. Mika, P.: Ontologies are us: A unified model of social networks and semantics. Web Semantics: Science, Services and Agents on the World Wide Web 5(1), 5–15 (2007)
24. Miller, G.A.: Wordnet: a lexical database for english. Commun. ACM 38(11), 39–41 (1995)
25. Pan, J.Z., Taylor, S., Thomas, E.: Reducing ambiguity in tagging systems with folksonomy search expansion. In: Aroyo, L., Traverso, P., Ciravegna, F., Cimiano, P., Heath, T., Hyvönen, E., Mizoguchi, R., Oren, E., Sabou, M., Simperl, E. (eds.) ESWC 2009. LNCS, vol. 5554, pp. 669–683. Springer, Heidelberg (2009)
26. Pass, G., Chowdhury, A., Torgeson, C.: A picture of search. In: The First International Conference on Scalable Information Systems (2006)
27. Rocchio, J.J.: Relevance feedback in information retrieval. In: Salton, G. (ed.) The SMART Retrieval System: Experiments in Automatic Document Processing. Prentice-Hall Series in Automatic Computation, ch. 14, pp. 313–323. Prentice-Hall, Englewood Cliffs (1971)
28. Salton, G., Buckley, C.: Improving retrieval performance by relevance feedback. In: Readings in Information Retrieval, pp. 355–364 (1997)
29. Schmitz, P.: Inducing Ontology from Flickr Tags. In: Proceedings of Collaborative Web Tagging, Workshop at WWW 2006 (May 2006)
30. Shah, C., Croft, W.B.: Evaluating high accuracy retrieval techniques. In: SIGIR 2004: Proceedings of the 27th Annual International ACM SIGIR Conference on Research and Development in Information Retrieval, pp. 2–9. ACM, New York (2004)
31. Shokouhi, M., Azzopardi, L., Thomas, P.: Effective query expansion for federated search. In: SIGIR 2009: Proceedings of the 32nd International ACM SIGIR Conference on Research and Development in Information Retrieval, pp. 427–434. ACM, New York (2009)
32. Sigurbjörnsson, B., van Zwol, R.: Flickr tag recommendation based on collective knowledge. In: WWW 2008: Proceeding of the 17th International Conference on World Wide Web, pp. 327–336. ACM, New York (2008)
33. Voorhees, E.M.: Query expansion using lexical-semantic relations. In: SIGIR 1994: Proceedings of the 17th Annual International ACM SIGIR Conference on Research and Development in Information Retrieval, pp. 61–69. Springer, New York (1994)
34. Wee, L.C., Hassan, S.: Exploiting Wikipedia for Directional Inferential Text Similarity. In: Fifth International Conference on Information Technology: New Generations, ITNG 2008, pp. 686–691 (April 2008)
35. Xu, J., Croft, W.B.: Query expansion using local and global document analysis. In: SIGIR 1996: Proceedings of the 19th Annual International ACM SIGIR Conference on Research and Development in Information Retrieval, pp. 4–11. ACM, New York (1996)
36. Yahia, S.A., Benedikt, M., Lakshmanan, L.V.S., Stoyanovich, J.: Efficient network aware search in collaborative tagging sites. In: Proceedings VLDB Endow., vol. 1, pp. 710–721. VLDB Endowment (2008)

37. Yin, Z., Shokouhi, M., Craswell, N.: Query expansion using external evidence. In: Boughanem, M., Berrut, C., Mothe, J., Soule-Dupuy, C. (eds.) ECIR 2009. LNCS, vol. 5478, pp. 362–374. Springer, Heidelberg (2009)
38. Zhou, D., Bian, J., Zheng, S., Zha, H., Giles, C.L.: Exploring social annotations for information retrieval. In: WWW 2008: Proceeding of the 17th International Conference on World Wide Web, pp. 715–724. ACM, New York (2008)

Ranking of Semantically Annotated Media Resources

Tobias Bürger[1] and Stefan Huber[2]

[1] Knowledge and Media Technologies Group, Salzburg Research, Salzburg, Austria
tobias@tobiasbuerger.com
[2] University of Applied Sciences Kufstein, Kufstein, Austria
mail@stefanhuber.at

Abstract. Ranking of resources is relevant in information retrieval systems to present optimally targeted results to initially issued queries. Traditional ranking approaches from the field of information retrieval are not always appropriate to be implemented in search engines based on semantics. Thus this paper proposes a customized ranking approach for multimedia resources based on semantic annotations which is based on the vector space model. An initial version of the ranking algorithm is evaluated and refinements are proposed which are again evaluated in an end user experiment.

1 Introduction

Retrieving media resources remains a challenge on the Web which is amongst others due to the inherent limitations of machine-driven multimedia understanding. One reason for that is the so-called Semantic Gap [9], which refers to the inherent gap between low-level multimedia features, which can be feasibly extracted automatically, and high-level features, which can not be accurately derived without human intervention. Describing multimedia resources through metadata is thus often seen as the only viable way to enable efficient multimedia retrieval. Semantic technologies have been identified as a potential solution to describe metadata in a machine processable way and by that eliminating difficulties in interpreting the content of media resources by search engines. The European project SALERO was concerned with the implementation of a semantic search engine based on semantic descriptions of media resources. This paper presents details on the search infrastructure of the so-called SALERO Semantic Workbench. It introduces its approach to rank search results based on their semantic annotations. It presents two iterations of the ranking algorithm and an evaluation of both.

The remainder of this paper is structured as follows: Section 2 gives an introduction to the SALERO project and its semantic search system. Section 3 presents the two versions of the ranking approach implemented in the search

T. Declerck et al. (Eds.): SAMT 2010, LNCS 6725, pp. 17–31, 2011.

system and Section 4 presents the evaluation of the second iteration of the algorithm. Section 5 presents the state of the art in the area of ranking of semantically annotated resources and Section 6, finally, provides a discussion of the results and concludes the paper.

2 Semantic Search and Annotation in the SALERO Semantic Workbench

In order to pave the way for the use of ontologies and semantic technologies in media production, SALERO developed a management framework for multimedia ontologies, tools to annotate existing media resources and semantic search facilities to retrieve resources based on semantic annotations (cf. [4, 14]). The SALERO Semantic Workbench supports the life cycle of multimedia ontologies, that is their creation, management, and use. This includes the following main functionalities:

- **Ontology Management.** whose central aspects include manual and semi-automatic creation of domain ontologies, alignment of different domain descriptions, translations of ontologies, versioning, or storage of ontologies.
- **Annotation Support.** whose central aspects includes the support for non-technological users with the annotation of media items which is realized in several annotation tools.
- **Semantic Search Support.** which offers advanced retrieval capabilities based on semantic annotations.

2.1 Annotation Support, Underlying Data, and Ontology

The SALERO workbench has been mainly designed for the retrieval of multimedia resources which are annotated with RDF triples. The annotation triples form a two level hierarchy: The upper level generally describes the content of the resource and in the lower level the general description can be specified. The annotation ontology used in SALERO formally describes the constituents of annotations and it additionally uses domain ontologies which define annotation concepts that are arranged in a taxonomy. Instances of the annotation concepts annotate the resources. Additionally to annotating a single resource, relationships between two resources can be created.

SALERO allows users to annotate resources using *annotation statements* that contain semantic elements which are defined in the SALERO annotation ontology.

Statements are in the form of < Concept isRelatedTo Concept 1 ... Concept n > which are triples where a concept can be set in relation to other concepts (cf. [13]).

2.2 Semantic Search Support

The workbench provides the so called Semantic Search service that supports two different types of queries: The first option are concept-based queries which are translated into SPARQL queries, the query language for RDF.[1] The second option are keyword-based queries which are executed in a search index which is generated based on the ontologies and instance data in the repository. The search index preserves the annotation triples and also stores subsumption information to enable query expansion. Keyword and concept queries can be mixed to increase the precision of keyword-based querying in the system. Furthermore selected WordNet relations of the concepts can be expanded to increase the result set while preserving precision.[2] Once the queries are answered, the results are ranked based on the semantics of the annotations attached to the returned results and their degree of match to the query (cf. Section 3).

3 Ranking of Media Resources in the Semantic Workbench

The ranking algorithm implemented in the semantic workbench has been developed in two iterations, both of which are described in the following.

3.1 Initial Ranking Approach

The initial ranking algorithm implemented in the semantic workbench follows the idea of the approach presented in [7] which is essentially based on the notion of similarity between a query document and documents in the result set. The approach has been designed to assess the relevance of annotations based on underlying domain ontologies and it takes into account the separation between instances and classes, the subsumption hierarchies in ontologies and part-of relations. It adapts the tf/idf weighting of feature vectors in the vector space model and, in analogy to the vectors, triples are used instead in this approach. Accordingly to tf/idf, the keywords from the query are transformed into triples and compared with the source triples annotating resources. Based on the differences of ontology based search to classical information retrieval (i.e the search for annotation instead of documents and the fact that documents are not represented by terms but by ontological annotations), the authors provided a set of idf-like measures such as *inverse class factor*, *inverse instance factor*, and *inverse triple factor* that account for the global usage of classes, individuals and triples in the annotation. *Inverse class* and *instance factor* are defined equally to the original idf metric. The *inverse triple factor* is defined as the total number of matching annotation triples compared to the total number of annotation triples. The *ontological instance relevance* is determined by the product of the *inverse instance factor* and the *inverse class factor*. The *ontological triple relevance* is determined based on the relevance of the subjects, the objects and, if the properties match, summed up with the *inverse triple factor*.

Intuitively, the *ontological triple relevance* is high for a given query if the subject and object of the document annotation triple are relevant given the

[1] http://www.w3.org/TR/rdf-sparql-query/

[2] http://wordnet.princeton.edu/

subject and object of the second document annotation triple, or the query triple respectively, and if the document annotation triple is rarely used. The annotation relevance is then the sum of the triple relevance divided through the number of triples used in the target annotation.

In the SALERO project, we implemented a slightly refined version of the explained approach. The implemented algorithm is illustrated in Listing 1.1 and explained thereafter:

```
0   rank(resource,query) {
1       score = 0.0;
2       normFactor = maxAnnotationsFromAllResources;
3       concepts = getConcepts(query);
4       n = resource.getNumberOfAnnotationConcepts();
5       for each (concept from concepts) {
6           nLowerLevel =
                resource.getNumberOfConceptLowerLevel(concept);
7           nTopLevel =
                resource.getNumberOfConceptTopLevel(concept);
8           conceptFrequency = (nTopLevel * 2 / n) +
                (nLowerLevel / n);
9           targetTriple =
                (resource.getAnnotationType(), related ,concept);
10          for each (sourceTriple from resource.getTriples()) {
11              itf = itf(targetTriple,sourceTriple);
12              icf =
                    icf(targetTriple.subject,sourceTriple.subject);
13              icf +=
                    icf(targetTriple.object,sourceTriple.object);
14              otr = itf * icf;
15              score += otr * concept.weight * conceptFrequency;
16          }
17      }
18      return score / normFactor;
```

Listing 1.1. Initial Ranking Approach

1. The concepts c and a weight denoting their relevance regarding to the user's keywords are retrieved from the search index (3). These concepts form the target concepts, which later are used to generate target triples.
2. For each target concept the concept frequency cf with respect to a resource is computed according to the following formula (4-8) whereas $i(c)$ denotes the number of instances of a concept either on the toplevel (tl) or lowerlevel (ll), and n the number of annotations of a resource:

$$cf(c) = \frac{i(c)_{ll} + i(c)_{tl} - 2}{n} \tag{1}$$

3. For each target concept (5) the ontological triple relevance (otr) (14) with respect to each source triple (s) is computed. Therefore a target triple (t) has

to be created, in which the subject is the annotation type of the resource, the predicate is "related to" and the object is the target concept (11).

$$otr(t, s) = (icf(t.subj, s.subj) + icf(t.obj, s.obj)) * itf(t, s) * multiplier \quad (2)$$

In order to compute the inverse class factor (icf) (12,13) and the inverse triple factor (itf) (13) a match function has to be introduced, which compares two concepts (a, b) and computes a matching value $(match())$. If two concepts are instances of the same class, they fully match and a matching value of 1.0 represents their matching degree. Otherwise, if the concepts do not fully match, but one concept is a super concept of the other, a partial match is achieved, which is represented by a matching value of 0.7. In the case of a complete mismatch, the matching value is 0.0. i denotes the total amount of instances and $i(a)$ the amount of instances of concept a

$$icf(a, b) = log(\frac{i}{i(a)}) * match(a, b) \quad (3)$$

The inverse class factor is higher for instances of a concept which is rarely used and where the target and source concept fully match.

The inverse triple factor (itf) is higher for rare triples. The multiplier boosts the ontological triple relevance by 1.5, if the subjects and the objects of both triples at least have a partial match, else it stays 1.0. Intuitively the ontological triple relevance is higher if the components of the triples match and the concepts or the triples are rarely used. o denotes the amount of all triples and $o(t)$ denotes the amount of triples with a similar structure to triple t.

$$itf(t) = log(\frac{o}{o(t)}) \quad (4)$$

4. The score (15) for a target concept c_{tar} is computed by summing up the ontological triple relevance for all source triples t_{src} with respect to the target triple multiplied by the concept frequency for the target concept and the target concept weight retrieved in step 1.

$$score = \sum_{\forall t_{src}} otr(t_{tar}, t_{src}) * icf(c_{tar}) * weight(c_{tar}) \quad (5)$$

5. The score (18) for the resource is finally divided by a normalization factor (2). The normalization factor is the amount of concepts of a resource, which has the most annotations.

3.2 Observed Weaknesses

During an initial user evaluation of the approach, various weaknesses could be observed. Some weaknesses are illustrated using the following example (cf.

Bing related to Telescope
Bing related to Smiling
Bong related to Smiling

Bing related to Boots
Bing related to Holding
Cold
Bing related to Scarf
Bong related to Earmuffs

Fig. 1. Exemplary Image With Annotations

Figure 1), while others are briefly discussed. We assume that if a search for "Bing telescope" is performed, it is very likely that the user intends to retrieve an image which contains the character Bing using a telescope. A RDF triple which would describe the user's information need best looks like: $< Bing >< relatedto >< Telescope >$. In the initial approach the algorithm ranks the image depicted on the right hand side in Figure 5 higher than the one on the left. However as the Triple $< Bing >< relatedto >< Telescope >$ is present in the left image and not on the right, it seems to be natural that the left image should be ranked higher in a search result than the right one.

The ineffectiveness of the initial ranking approach has various reasons, which are discussed in the following: The target triples (10) which are generated based on the concepts reflecting the query are missing important triples. As described previously, the triples are built according to the structure: $< annotationtype >< relatedto >< concept >$. This process leaves out important triples, such as the one which describes the user's information need best from the above mentioned example: $< Bing >< relatedto >< telescope >$. In the initial ranking approach triples of the form $< targetconceptx >< relatedto >< targetconcepty >$ are not influencing the score of a resource, but should do according to their highly perceived importance. The concept frequency (5-9), which is computed in step 2, influences the score in a contradictory manner. In the example, the concept "Bing" has a high frequency in the right image and additionally is in the top level of the annotation hierarchy and therefore has doubled influence. In contrast to that, the left image includes both concepts for which the search is performed in the example, but the concept "telescope" resides in the lower level of the annotation hierarchy and therefore has less influence. Intuitively, the computation of the concept frequency represents the structure of the image annotation correctly, but does not acknowledge the fact that the annotations are represented as triples. Therefore resources for whose annotations certain concepts have a high frequency in the top level boost the score to such a significant extent, that outweighs the score of those resources which contain more relevant concepts, but in lower frequency and in the lower position in the hierarchy. As mentioned in step 4, for each source triple a part of the whole score is computed by multiplying the ontological triple relevance with the concept frequency and the concept weight (16). For source triples which occur multiple times this score is computed multiple times as well. In principle, it would be enough to let it influence the score only once. It is anyway multiplied with the concept frequency and therefore a multiple occurrence of a concept is already taken into consideration. Finally there has not been any consideration of the amount of target triples which match against the source triple of a resource. Resources matching fewer target triples should be ranked lower, than resources matching a higher amount of triples.

3.3 Refined Ranking Approach

According to the observed weaknesses of the initial ranking approach, several refinements have been implemented. The refinements are presented in the following, accompanied by a motivating example. In parenthesis a reference to the line number of the pseudo code from Listing 1.2 is given.

We assume that wrt. a query "Bing umbrella" the right image in Figure 2 is more relevant and should be ranked higher than the left one. With the initial ranking approach this could not be achieved.

Bing related to Holding
Bing related to Ice cream
Bong related to Smiling
Umbrella

Bing related to Waving
Bong related to Smiling
Rock
Bing related to Umbrella

Fig. 2. Exemplary Image With Annotations 2

```
0  rank (resource ,query){
1    score = 0.0, tripleHits = 0.0, tripleBoost = 0.0;
2    normFactor = maxAnnotationsFromAllResources;
3    concepts = getConcepts(query);
4    targetTriples = getTargetTriplesFromConcepts(concepts);
5    sourceTriples = resource.getTriples();
6    for each (targetTriple from targetTriples) {
7        tripleBoost = 1.0;
8        if (sourceTriples.contain(targetTriple) {
9           if (targetTriple.subject ==
               resource.annotationType) {
10              tripleBoost +=
                   log(targetTriples.count(targetTriple) /
                   targetTriples.size());
11              tripleHits++;
12          } else {
13              tripleBoost += targetTriples.count(targetTriple)
                   / targetTriples.size() * 1.5;
14              tripleHits += 1.5;
15 }
16          for each (sourceTriple from
               resource.getDistinctTriples()) {
17             itf = otr(targetTriple,sourceTriple);
18             icf =
                  icf(targetTriple.subject,sourceTriple.subject);
19             icf +=
                  icf(targetTriple.object,sourceTriple.object);
20             score += itf * icf * triple.weight * tripleBoost;
                  } } }
21   return score * (tripleHits / targetTriples.size()) /
         normFactor;
```

Listing 1.2. Refined Ranking Approach

The first refinement changes the way how target triples are computed. In the initial approach, each retrieved target concept from the keywords is turned into a triple according to the following schema: $< annotationtype >< relatedto >< targetconcept >$. The important triples which would represent the user's information need best, are completely left out (i.e., $< Bing >< relatedto >< umbrella >$). Therefore the process of creating target triples is augmented by taking two different target concepts as either subject or object (4,5). The query of the example "Bing umbrella" leads to the following target triples in the refined approach: $< Bing >< relatedto >< umbrella >$, $< umbrella >< relatedto >< Bing >$, $< imageannotation >< relatedto >< Bing >$, $< imageannotation >< relatedto >< umbrella >$. The second refinement substitutes the concept frequency with a *triple boost factor* (*tbf*; 10,13). As in the initial approach the concept frequency is not considering the fact that annotations are formed by triples, this substitution is made. Triples which are of the form $< annotationtype >< relatedto >< concept >$ are of lower importance than triples of the form $< conceptx >< relatedto >< concepty >$ as the latter one describes a user's information need better. If a target triple matches the form $< annotationtype >< relatedto >< concept >$ the *triple boost factor* is computed according to the following formula in which t_{tar} denotes the target triple, $n_{tt}(r)$ the amount of target triples of a resource r and $n_{at}(r)$ the amount of annotation triples:

$$tbf(t_{tar}) = \sqrt{1 + \frac{n_{tt}(r)}{n_{at}(r)}} \tag{6}$$

As the *triple boost factor* should always increase the score, it should always be bigger than one. Therefore the ratio of target triples in the annotations of the resource to all triples annotating the resource is added to 1. Further the square root is taken to lower the influence of triples of this form. The factor for the more important triples of the form $< conceptx >< relatedto >< concepty >$ is computed according to the following formula:

$$tbf(t_{tar}) = 1 + \frac{n_{tt}(r)}{n_{at}(r)} * 2.0 \tag{7}$$

To compute the *triple boost factor* for triples of the latter form, the square root is not taken and further the frequency is multiplied by 2.0 to accent the importance of these triples. Again, the factor should always increase the score and therefore it should be bigger than one, thus 1 is added. Table 1 exemplarily presents the results of the triple boost factor computation for the above mentioned example.

The third refinement refers to step 4 of the initial ranking algorithm. In the initial approach a score is computed for each source triple wrt. to the current target triple. For similar source triples the score is computed as often as the source triple occurs. In the refined approach, the score is computed only once, regardless if the source triple occurs multiple times (16). This aspect is changed in the refined approach because the frequency of occurrence is already taken into account via the *concept frequency* in the initial approach and via the *triple boost factor* in the refined approach.

The fourth refinement introduces a *triple hit factor*, which is determined by the amount of matching target triples against a resource's source triples. For each triple hit of the form $< annotationtype >< relatedto >< concept >$ the factor is augmented by 1.0 (12). For a triple hit of the form $< conceptx >< relatedto >< concepty >$ the factor is augmented by 1.5 (14). The final score of

Table 1. Triple Boost Factor Computation

triples	left image	right image
$< Bing >< relatedto >< umbrella >$	not computed	1.2857
$< umbrella >< relatedto >< Bing >$	not computed	not computed
$< imageannotation >< relatedto >< Bing >$	1.1338	1.1338
$< imageannotation >< relatedto >< umbrella >$	1.0690	not computed

Table 2. Triple Hit Factor Computation

	left image	right image
triple hit factor	$\frac{2}{4} = 0.5$	$\frac{2.4}{4} = 0.625$

a resource is computed according to the following formula (21) in which nm_{tt} denotes the amount of matching target triples and n_{tt} the amount of target triples:

$$score_{final} = score * \frac{nm_{tt}}{n_{tt}} \tag{8}$$

Table 2 shows an exemplary computation of the *triple hit factor* for the above mentioned example.

4 Evaluation

In this section the evaluation of the refined ranking approach is including the setting of the experiment and its results.

Experiment. For the evaluation of the refinements of the ranking algorithm an end-user experiment has been performed. Therefore a panel of five test users was chosen to play the role of a ranking algorithm. In three runs each test user had to order a set of resources according to their relevance to a query. To perform the experiment, images from Tiny Planets[3] were chosen and their annotations were presented to the test users on a sheet of paper. To perform the experiment no extensive domain knowledge was needed. In a short conversation the experiment was explained to the participants beforehand. The main focus of this introduction was the structure of the annotations and the underlying ontology, further the domain of Tiny Planets was briefly introduced. There was no time limit set for the completion of the experiment. In the first run, the query "Bing telescope" was given to the test user together with 8 images. In the second run the query "Alien smiling" was given to the test users together with seven images. Finally in the third run the query "Bing umbrella" was given to the test users and six images. In order to evaluate the implemented approaches, the Spearman rank correlation coefficients has been used. For each result of the test users a correlation between the refined approach and the initial approach has been computed. The correlation is a value between the interval -1 and 1, whereas 1 denotes a complete correlation and -1 means one ranking is the reverse of the other. With this measure the effectiveness of the two ranking algorithms can be compared. If the correlation for the refined approach is higher than the one for the initial approach, the refined one is better. If the correlation for the refined approach is lower than the one for the initial approach, the initial approach is better. If the correlation for the two approaches is the same, they are equal.

[3] http://www.tinyplanets.com/

Results. Tables 3 – 5 present the results of the experiment for different queries. On the left hand side of each table the resources are listed represented using capital letters. In the subsequent columns the ranking position of a resource computed by one of the two algorithms and by the test persons is shown. Ranks are consequently numbered from 1 to 7, whereas 1 represents the highest position. The last two rows show the correlations of the initial algorithm with all rankings of the test persons and the correlations of the refined approach with all the test manually-derived rankings. It can be easily observed that in all cases the correlation of the refined approach is higher than the correlation of the initial approach. According to the results of the experiment the refinements have improved the algorithm.

Table 3. Query 1 "Bing Telescope"

resources	refined approach	initial approach	P1	P2	P3	P4	P5
A	1	1	2	2	2	2	3
B	8	7	8	8	7	6	8
C	3	8	3	3	3	4	2
D	5	2	4	7	5	3	6
E	4	3	5	5	6	8	7
F	6	5	6	4	4	5	4
G	7	6	7	6	8	7	5
H	2	4	1	1	1	1	1
correlation refined approach			0.952	0.857	0.857	0.667	0.714
correlation initial approach			0.452	0.214	0.310	0.357	0.000

Table 4. Query 2 "Alien Smiling"

resources	refined approach	initial approach	P1	P2	P3	P4	P5
A	5	4	4	5	4	4	5
B	4	1	5	4	5	5	4
C	1	2	1	1	1	1	1
D	7	7	6	6	6	7	7
E	3	3	3	3	3	3	3
F	6	6	7	7	7	6	6
G	2	5	2	2	2	2	2
correlation refined approach			0.929	0.964	0.929	0.964	1.000
correlation initial approach			0.500	0.607	0.500	0.536	0.643

Table 5. Query 3 "Bing Umbrella"

resources	refined approach	initial approach	P1	P2	P3	P4	P5
A	2	2	3	2	1	2	2
B	3	3	2	3	3	3	3
C	1	6	1	1	2	1	1
D	5	5	4	4	4	6	5
E	4	1	6	5	6	4	6
F	6	4	5	6	5	5	4
correlation refined approach			0.771	0.943	0.771	0.943	0.771
correlation initial approach			-0.543	-0.314	-0.257	-0.029	-0.429

5 Ranking of Semantically Annotated Resources: State of the Art

Many recent approaches exploit ontologies to improve retrieval of information and to improve precision and recall by inferring implicit facts. Many of these approaches concentrate on strict boolean querying of knowledge-bases or implement existing approaches from the Information Retrieval literature for text-based collections rather than assessing relevance for the retrieved documents of facts from the knowledge-base. Some approaches rank results based on different aspects which we present in the following.

5.1 Ranking Based on Topic-Similarity

Approaches such as Bibster [3] allow searching for resources by topic and rank the results according to their similarity to the requested topic. Bibster devised a similarity function between a topic and a resource description. The similarity is computed based on the closeness of the topics in the topic hierarchy given by the used ontologies.

Another approach which ranks documents based on the similarity of query terms and document terms is applied in Ontobroker [10]. There, similarity is determined based on the relative place of two terms in a common hierarchy. The similarity value equals 1 if the objects coincide, 0 if two concepts do not share a common hierarchy and is otherwise determined based on the distance of terms in the hierarchy.

5.2 Preference-Based Ranking

Preference based ranking has been investigated in [8]. There existing query languages for RDF are extended with preference boolean operators which allow users to influence the ranking of results. The ranking of results follows a theory put forward by Chomicki [5] and implements a qualitative approach to preferences. This approach allows treating preferences independently and results in a partial preference order of the results. The authors introduce a relational algebra, an extension to the SPARQL query language supporting this algebra, and finally an implementation of the extension. So instead of best matches in all query dimensions their approach allows to weight between the results in the dimensions. An approach with a similar intention has been introduced in the Semantic Web Services area by Toma et al. [12]. There, users are able to query for Semantic Web Services using a set of so-called nonfunctional properties (nfp) of a service and preferences for these nfp's. The users can specify which preference should be used for ordering and should declare the importance of the nfp. These preferences are then used for ranking the results. Their approach takes multiple nfp's (criteria) into account but does not exploit the semantics of ontologies used for retrieval. The authors have further extended this approach to include social data and by that measure the popularity of a Web service [11].

5.3 Relationship Ranking

Relationship ranking approaches rank results based on the measurement of relationships. In [1], for istance, the authors introduce a system to rank complex relationships, i.e. so called semantic associations which are path sequences consisting of nodes and edges connecting two entities. The intention of this approach is to rank results for queries which involve two entities using semantic and statistical metrics. Whereas semantic metrics are based on the aspects of the involved ontologies, statistical metrics are based on statistical aspects such as the degree of connectivity of entities and relationships. The important difference to other approaches is that this approach ranks semantic associations and not documents which could, however, be used to influence ranking of documents as well. The semantic metrics include the ontological context, which captures regions in ontologies to which a user can assign weights, subsumption hierarchies, and trust values. Statistical metrics include rarity of occurring entities and relationships, popularity of entities and the length of associations.

Another approach to rank based on relationships is introduced in [6]. Here the authors present an algorithm which is based on spread activation techniques and which can be used to find related concepts in an ontology. The difference to the previous approach is, that not only subsumption hierarchies are exploited to find related concepts, but also arbitrary relations between concepts. In contrast to traditional searches their approach retrieves concepts even if the keyword does not match the concept. This approach applies different measures to estimate the degree of relatedness of two concepts, e.g., the cluster measure and the specifity measure. The former indicates the similarity between two concept instances. The latter measure captures the specifity of the relation by applying an Inverse Document Frequency (idf) metric to ontologies. Subsequently they apply a combination of both measures. Based on these measures, and an initial spreading activation value, related concepts are retrieved.

Another relationship ranking approach is SemRank [2]. SemRank aims to determine the relative importance of relationships found with respect to a user's context. This means that it tries to rank results that a user most likely expects on top positions using metrics that measure the predictability of these results. Factors to measure the predictability of a result include the specifity of a result and discrepancies in the structure of the result based on the query schema. This means that a commonly occurring relation is more predictable than a rarely occurring one. Finally, relevance is further assessed based on the semantic matches of keywords given by end users.

5.4 Ontology-Based Ranking

Other approaches for ranking of knowledge in the Semantic Web take the whole knowledge model into account, including the one described in [10]. In this approach the relevance of a found resource is determined based on the level of relation instances whereas the relevance of an answer will be determined based on which are combined. The relevance of a relation instance is determined based on its ambiguity and specifity. The relevance of a node in the tree depends on the relevance of its child nodes. Thus the relevance is propagated in the tree. Most notably every relation is treated differently which makes this approach appropriate if different relations are present and if these relations should have a different influence on the ranking. Another approach which exploits information from domain ontologies to perform ranking is the approach presented in [7] on which the algorithm presented in this paper is based on (cf. Section 3).

6 Discussion and Conclusions

First, it has to be mentioned that a ranking of semantically annotated resources can only be as good as the annotations provided. If the annotations of a multimedia resource are not representing the content correctly, then the ranking cannot be effective. Second, the experimental evaluation of the refinements presented in the previous section cannot prove the increase of effectiveness regarding to the initial approach as a panel of five humans is typically not a critical mass. However the experimental results show a clear tendency that the results are ranked closer to a user's expectations than in the initial approach. As an information retrieval system can only retrieve the relative best answers to a query, a ranking algorithm can never be perfect. Currently there is already one aspect of the underlying resource annotations which is not taken into consideration by the algorithm. As briefly mentioned in Section 3, relationships between resources can be created. Therefore if a resource is popular, which means it has many in- and outgoing relationships it could be more relevant in some situations. As we have not yet found any evidence for this assumption, it has not been considered.

In the future, we aim to deploy the ranking algorithm presented in this paper in a different semantic search engine and potentially evaluate it with a larger set of end users interactively in the deployed search engine.

Acknowledgments. The research leading to this paper was partially supported by the European Commission under contract IST-FP6-027122 "SALERO".

References

1. Aleman-Meza, B., Halaschek-Wiener, C., Arpinar, I.B., Ramakrishnan, C., Sheth, A.P.: Ranking complex relationships on the semantic web. IEEE Internet Computing 9(3) (2005)
2. Anyanwu, K., Maduko, A., Sheth, A.: Semrank: ranking complex relationship search results on the semantic web. In: Proceedings of the 14th International Conference on World Wide Web, Chiba, Japan, May 10-14 (2005)
3. Broekstra, J., Ehrig, M., Haase, P., Van Harmelen, F., Mika, P., Schnizler, B., Siebes, R.: Bibster – A semantics-based bibliographic peer-to-peer system. In: McIlraith, S.A., Plexousakis, D., van Harmelen, F. (eds.) ISWC 2004. LNCS, vol. 3298, pp. 122–136. Springer, Heidelberg (2004)
4. Bürger, T.: Multimedia ontology life cycle management with the salero semantic workbench. In: Proceedings of the Workshop on Semantic Multimedia Database Technologies, SeMuDaTe 2009 (2009)
5. Chomicki, J.: Preference formulas in relational queries. ACM Trans. Database Syst. 28(4), 427–466 (2003)
6. Rocha, C., Schwabe, D., Aragao, M.P.: A hybrid approach for searching in the semantic web. In: Proceedings of the 13th International Conference on World Wide Web, New York, NY, USA, May 17-20, pp. 374–383 (2004)
7. Ruotsalo, T., Hyvnen, E.: A method for determining ontology-based semantic relevance. In: Wagner, R., Revell, N., Pernul, G. (eds.) DEXA 2007. LNCS, vol. 4653, pp. 680–688. Springer, Heidelberg (2007)
8. Siberski, W., Pan, J.Z., Thaden, U.: Querying the semantic web with preferences. In: Cruz, I., Decker, S., Allemang, D., Preist, C., Schwabe, D., Mika, P., Uschold, M., Aroyo, L.M. (eds.) ISWC 2006. LNCS, vol. 4273, pp. 612–624. Springer, Heidelberg (2006)
9. Smeulders, A., Worring, M., Santini, S., Gupta, A., Jain, R.: Content-based image retrieval at the end of the early years. IEEE Trans. Pattern Anal. Mach. Intell. 22(12) (2000)
10. Stojanovic, N., Studer, R., Stojanovic, L.: An approach for the ranking of query results in the semantic web. In: Fensel, D., Sycara, K., Mylopoulos, J. (eds.) ISWC 2003. LNCS, vol. 2870, pp. 500–516. Springer, Heidelberg (2003)
11. Toma, I., Ding, Y., Chalermsook, K., Simperl, E., Fensel, D.: Utilizing web 2.0 in web service ranking. In: Proceedings of the 3rd International Conference on Digital Society, ICDS 2009 (2009)
12. Toma, I., Roman, D., Fensel, D., Sapkota, B., Gomez, J.M.: A multi-criteria service ranking approach based on non-functional properties rules evaluation. In: Krämer, B.J., Lin, K.-J., Narasimhan, P. (eds.) ICSOC 2007. LNCS, vol. 4749, pp. 435–441. Springer, Heidelberg (2007)
13. Weiss, W., Bürger, T., Villa, R., Punitha, P., Halb, W.: Statement-based semantic annotation of media resources. In: Proceedings of the 4th International Conference on Semantic and Digital Media Technology (SAMT), Graz, Austria, pp. 02.12–04.12 (2009)
14. Weiss, W., Bürger, T., Villa, R., Swamy, P., Halb, W.: Salero intelligent media annotation & search. In: Proceedings of the International Conference on Semantic Systems, I-Semantics 2009 (2009)

Enabling Semantic Search in a News Production Environment

Pedro Debevere[1], Davy Van Deursen[1], Dieter Van Rijsselbergen[1], Erik Mannens[1], Mike Matton[2], Robbie De Sutter[2], and Rik Van de Walle[1]

[1] Ghent University - IBBT, Multimedia Lab, Belgium
{pedro.debevere,davy.vandeursen,dieter.vanrijsselbergen, rik.vandewalle}@ugent.be
[2] VRT-medialab, Belgium
{mike.matton,robbie.desutter}@vrt.be

Abstract. News production is characterized by a complex and dynamic workflow, in which it is important to produce and broadcast reliable news as fast as possible. In this process, the efficient retrieval of previously broadcasted news items is important, both for gathering background information and for reuse of footage in new reports. This paper discusses how the quality of descriptive metadata of news items can be optimized, by collecting data generated during news production. Starting from a description of the news production process of the Flemish public service broadcaster in Belgium (VRT), information systems containing valuable metadata are identified. Subsequently, we present a data model that uniformly represents the available information generated during news production. This data model is then implemented using Semantic Web technologies. Further, we describe how other valuable data sets, present in the Semantic Web, are connected to the data model, enabling semantic search operations.

Keywords: News item retrieval, News production, Semantic Web.

1 Introduction

Efficient media search applications can highly improve productivity in various domains. An important requirement for efficient media retrieval is the availability of high quality metadata documenting the archived media [5,6,8,16]. This is also the case within a news production environment, where professional archive users spend considerable amounts of time searching in the media archive in order to find useful media for reuse in news broadcasts [14]. However, in a news production environment it is also important to produce and distribute news as soon as possible to as many channels as possible, in an audiovisual quality as good as possible [9]. Therefore, metadata generation is often reduced to an absolute minimum during news production [12]. Consequently, dedicated archivists are responsible for the generation of high quality metadata as a last step in the news production chain in order to facilitate efficient media retrieval.

T. Declerck et al. (Eds.): SAMT 2010, LNCS 6725, pp. 32–47, 2011.

In this paper, we investigate the news production process of the Flemish public service broadcaster in Belgium (i.e., Vlaamse Radio- en Televisieomroep (VRT[1])). In large news production enterprises, such as VRT, several databases and information systems (e.g. subtitling and rundown management systems) are used during the news production process, each containing valuable information about the news and related media being produced. These information sources are often not or barely coupled. In addition, every information source has its own information storage facility resulting in a non-uniform data representation. However, most information contained in these information systems can often serve as valuable contributors of metadata describing the archived audiovisual material.

Therefore, the 'MediaLoep' project[2] investigates how news-related audio and audiovisual content can be found in a more effective, efficient, and easy way. One of the goals of this research project is to increase the amount of quality metadata by capturing and structuring the available information during news production. This automatically retrieved metadata can then be used by archivists as a starting point for further enrichment, leading to faster availability and more accurate search results, which improves the overall efficiency and productivity of news program editors.

As a first step, we investigate VRT's news production process and identify information sources containing valuable metadata. Subsequently, a data model is developed which covers and represents the information present in the various identified information sources in a uniform way. The model is implemented using Semantic Web technologies, which enable a formal and machine-understandable representation of the metadata. Further, the identified information sources need to be mapped to our data model. Finally, in order to enrich our metadata even more, we elaborate on the potential of connecting information stored according to the data model with other (external) data sets, which are available in the Semantic Web [3].

2 News Production Process

The following subsections describe VRT's news production process in order to facilitate the identification of valuable data sources for later retrieval of audiovisual content. This process is also illustrated in Fig. 1, where important steps in the news production process are labeled and referred to in the following subsections.

2.1 Editorial Planning and Management

The main tool used for news production at VRT is Avid's iNEWS[3]. This tool is used by directors and editors to create and manage news rundowns. A rundown consists of a list of items that will be covered during a news broadcast. An item

[1] http://www.vrt.be
[2] http://www.vrtmedialab.be/index.php/english/project/medialoep/
[3] http://www.avid.com/US/products/inews/

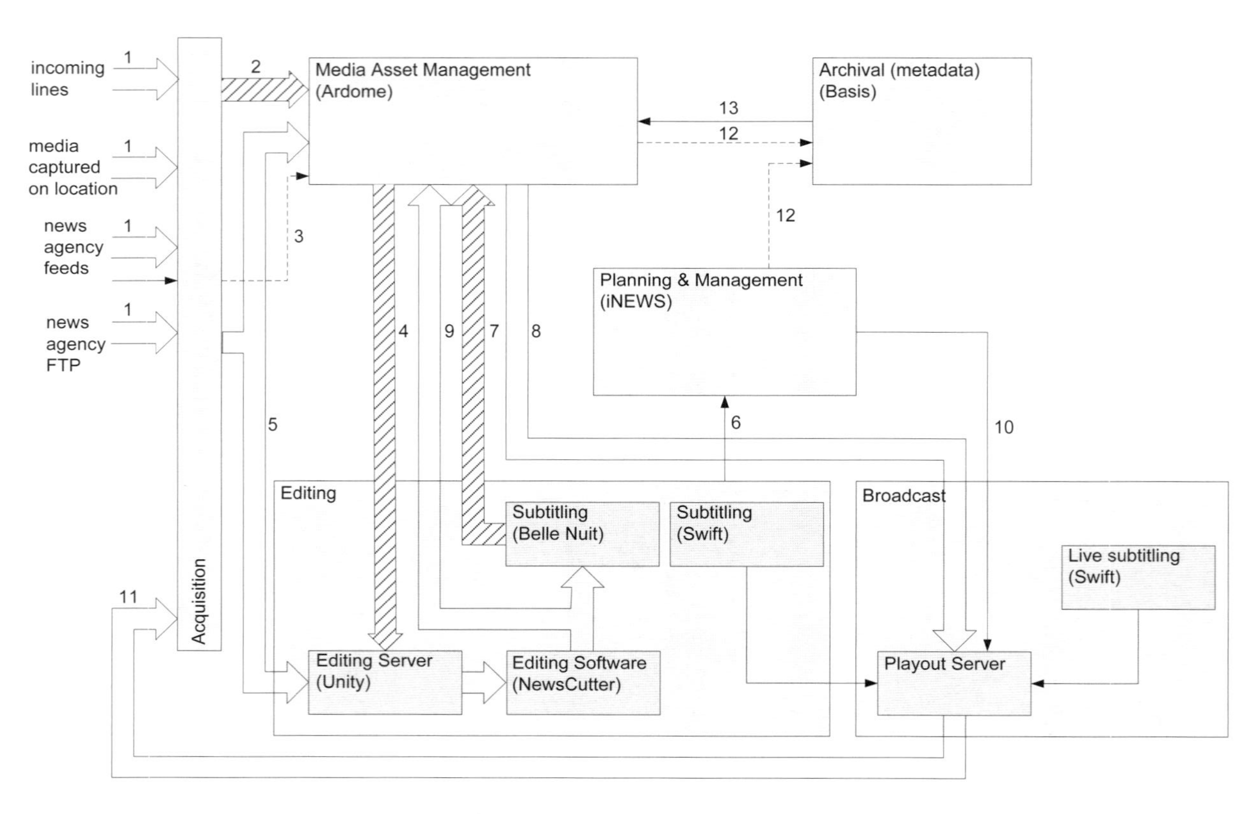

Fig. 1. News production process

is an atomic unit of a news rundown, e.g. a live interview, a report made by a journalist, or an introductory text displayed on the autocue[4] and read by the news anchor.

2.2 Acquisition

For most planned news items, relevant audiovisual content is needed. There are multiple options to obtain content related to a news item (label (1) in Fig. 1):

- audiovisual content captured by a news crew on location (typical for national news);
- content from incoming lines shot by other news providers (e.g., Reuters, EBU Eurovision and APTN);
- news feeds from news agencies, such as EBU Eurovision and Reuters, containing media files and additional metadata (typically represented according to the NewsML-G2 standard [10]);
- reuse of suitable audiovisual material obtained from the archive.

2.3 Ingest and Storage

Captured media is ingested and stored on servers managed by Ardome (Vizrt)[5]. Ardome is a Media Asset Management (MAM) system and is one of the core components in VRT's media production process. Consequently, it has many links with other information systems. Ardome contains all the produced media resources, both unfinished material as well as finished footage. In addition to media storage, Ardome provides other functionalities such as browsing, rough cut editing, and searching. In order to facilitate browsing and searching, low resolution versions of stored media are generated and accompanying metadata such as the title, audio track information and episode number can be inserted. However, during ingest, metadata insertion is kept to a minimum as this is currently a manual operation and consequently takes too much time. Therefore, metadata is often limited to a filename and a path indicating the location where the media is to be stored. Because captured content often needs further editing, content is also ingested in an editing server (Avid Unity ISIS[6]), which contains all media to be edited. As the editing server only hosts rough content that needs editing, the lifetime of this media is restricted to 72 hours.

In order to save time, content from incoming lines is often simultaneously ingested in the MAM and the editing server, resulting in a dual ingest (label (5) in Fig. 1). If dual ingest is not possible, the media is first ingested in the MAM (label (2) in Fig. 1) and is then sent from the MAM to the editing server (label (4) in Fig. 1). When content is captured from news feeds, the additional metadata is also inserted in the MAM (label (3) in Fig. 1). Note that the latter is again a manual operation (which is also indicated by a dashed line in Fig. 1).

[4] An autocue, also known as teleprompter, is a display device used to provide the news anchor with a continuous text feed during broadcast.

[5] `http://vizrt.com/products/article138.ece`

[6] `http://www.broadcastautomation.com/products/unityISIS/index.asp`

2.4 Editing

After ingest, editing can be performed. The journalist who has been given the task to cover a news item retrieves the corresponding media from the editing server and starts editing it using Avid NewsCutter[7]. Simultaneously, anchor text (appearing on the autocue during broadcast) is provided in iNEWS by the journalist which afterwards is reviewed by the news anchor. Textual information that must appear as graphics on screen during the broadcast of a news item (e.g., the name of the interviewed person) is also provided in iNEWS (label (6) in Fig. 1).

An important task during editing is the generation of subtitles. Two types of subtitles can be considered. The first type, referred to as open subtitles, are subtitles that are 'burned' into the picture and therefore always appear on screen. A typical example of the use of open subtitles is when a translation is needed for an interviewed person speaking a foreign language. The second type of subtitles, referred to as closed subtitles, are by default not displayed but can be retrieved when requested (e.g., via Teletext).

A journalist is responsible for creating open subtitles and inserting these in iNEWS. A copy of the edited media is rendered containing the open subtitles on the screen using the Belle Nuit tool[8]. The edited media with subtitles is transferred to the MAM (label (7) in Fig. 1) and afterwards sent to the playout server for broadcasting (label (8) in Fig. 1). Also a link is provided in iNEWS relating the edited media with the corresponding item in the rundown. A version without subtitles is also sent to the MAM for later archiving (label (9) in Fig. 1).

If there is time left, journalists also create the closed subtitles using Swift[9]. If this is not the case, live subtitling is performed during broadcast by the subtitling department.

2.5 Broadcast

During broadcast, the edited media (with open subtitles) is available on the playout server. iNEWS executes the news rundown (label (10) in Fig. 1) and sends the anchor text to the autocue. Textual information (e.g., the name of the interviewed person) is also displayed on screen when needed using the Character Generator (CG)[10].

The broadcast is again captured on an incoming line (label (11) in Fig. 1) by the media management department in order to have a copy of the integral news broadcast (which includes the open subtitles and textual information displayed by the CG on the screen). This captured broadcast is then also marked for archival.

[7] `http://www.avid.com/US/products/NewsCutter-Software/index.asp`
[8] `http://www.belle-nuit.com/subtitler/index.html`
[9] `http://www.softelgroup.com/product_info_1.aspx?id=0:53799&id=0:53783`
[10] A character generator is a device or software used to superimpose text onto video.

2.6 Archival

Archival is the last step in the news production process. An archivist who has been given the task to archive a news broadcast retrieves the rundown of the news broadcast from iNEWS and then searches for the corresponding media fragments in the MAM (label (12) in Fig. 1). Note that this is a manual operation and therefore indicated as a dashed line in Fig. 1. If a related fragment is found, the archivist watches it and generates a record containing relevant metadata using Open Text's Basis. Basis is the main tool used to document archived media at VRT in order to facilitate media retrieval. It is also the main tool for media search, e.g., when searching for archived content for reuse. Typically, a user searches for a relevant media fragment in Basis and subsequently retrieves it from the MAM.

As the archival step is performed as the last one in the media production process, it typically takes a few days before a media fragment is documented in Basis. When a Basis record is generated, some metadata fields that are also present in the MAM are transferred from Basis to the MAM (label (13) in Fig. 1), overwriting previously entered metadata in the MAM (e.g., metadata taken from the NewsML-G2 documents).

3 Production Process Evaluation

As can be seen from the discussion of the news production process, data is generated and spread across the entire production chain. Unfortunately, during a media search operation, only information generated by an archivist at the end of this chain is currently used. Other, possibly valuable information, is not used when searching for relevant media. We identified the following information sources containing valuable additional information for media search:

- **Ardome:** Ardome is the central MAM system used by VRT, containing all produced media. During ingest, metadata such as title, episode number, descriptive information of the audio tracks, aspect ratio, and video format can be inserted. However, as already noted in the previous section, this manual metadata insertion is always limited due to time constraints. Once the media has been documented in Basis, metadata is copied from Basis to Ardome.
- **News Agency provided metadata:** Incoming media from news agencies, such as EBU Eurovision and Reuters, is accompanied by metadata documenting the media in the NewsML-G2 format. This metadata can be inserted into the MAM during ingest. Typical metadata contained in a NewsML-G2 document are titles and textual descriptions (in English) of the media content.
- **iNEWS:** iNEWS contains important information as it is VRT's main tool for managing news productions. It is used to create the entire rundown of a news broadcast. Every news item from the rundown can be provided with anchor text, open subtitles, and textual information that must appear on the screen during broadcast by the CG.

- **Swift:** Closed subtitles are generated using the Swift tool. In addition to the subtitle text, Swift contains subtitle layout, spoken language indication, and timestamps indicating the appearance and disappearance time of a subtitle.
- **Basis:** Basis is considered to be the main tool for annotating and describing media resources at VRT. Basis has a relational database structure and currently contains over 600 000 records documenting media fragments spanning a period of over 20 years. A Basis record defines metadata fields such as title, duration, keywords, textual description, journalist, and identification number of the corresponding media in the MAM. Some fields, for example the keywords and journalist fields, can only contain controlled values, defined by manually maintained thesauri. The keywords thesaurus is the largest and contains over 300 000 terms. In addition to defining terms, relationships between terms are defined indicating 'narrower than', 'related' and 'used for'-relationships. Unfortunately, keywords are not categorized in terms of types, such as persons or locations.

By unifying all the information generated during news production, more efficient search applications can be realized. Also, by capturing information generated during news production, metadata can be generated and already filled in when creating an archival record, allowing an archivist to focus on the further enrichment of this metadata.

Subtitle information can significantly improve the media search operation [5,6]. For example, when a certain keyword appears in a subtitle, timestamp information can be used to start playback from the indicated timestamp. Closed subtitles, generated with the Swift tool, are provided with such timing information. Unfortunately, the appearance and disappearance timestamps for open subtitles are not available, as journalists use the Belle Nuit tool only to paste the subtitles onto the relevant frames and no related timing information is stored during this operation. However, this timing information can be obtained after media production through the use of speech analysis or optical character recognition (OCR) tools.

Timing information indicating the appearance time of CG text can also improve media search operations. For example, when a user searches for media related to a person, timing information related to the CG text containing the name of this person can be used to start playback at this point. However, the exact moment CG text is displayed is decided during live broadcast and this information is currently not stored. This timing information could be reconstructed through the application of feature extraction tools on the generated media.

It is clear that although much of the data generated during the news production process can be used for media search operations, not all this information is currently stored, and therefore introduces the need for post-production operations to reconstruct this information. In order to avoid the need for these costly post-production operations, production systems should be changed in order to store this information.

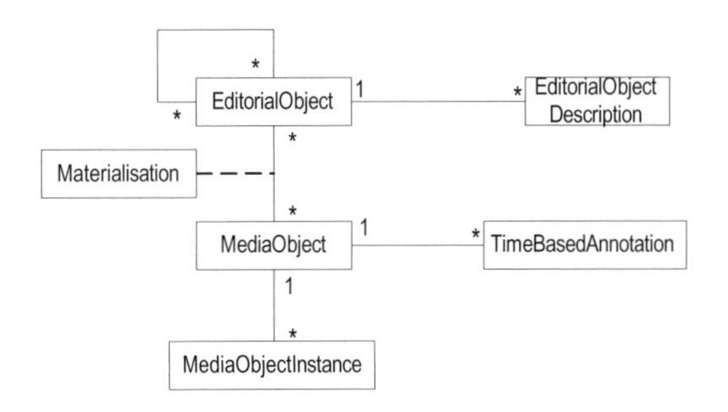

Fig. 2. Core concepts of the PISA media production model

In addition to the fact that not all information is stored, the current news production process does also not guarantee the preservation of inserted metadata. For example, when a news item from a news agency is ingested, a manual insertion of metadata provided in the accompanying NewsML-G2 document is performed into the MAM. However, when an archival record is generated in Basis documenting the archived media, an update operation is performed which overwrites the original metadata inserted into the MAM. This introduces the loss of the metadata obtained from the NewsML-G2 document although this metadata might be important when searching for media and should therefore be preserved too.

4 Integrating Information Sources: Data Model

In order to represent the information present in the various identified information sources in a uniform way, we propose a data model specifically designed to preserve metadata along the news production process. The data model is based on the PISA data model [17] that was developed as part of the IBBT PISA research project[11]. Note that this model is also compatible with the P/Meta data model [7] defined by the European Broadcasting Union (EBU) and shares business objects with the BBC SMEF model [1]. The core concepts of the PISA data model are illustrated in Fig. 2.

The PISA data model centers around four elemental types of processes that contribute to media production, as described in [17]. The lifecycle of an entire product (e.g, a news broadcast) starts with the product planning process. During product planning, initial product requirements are specified. Because this information is typically delivered from an external Enterprise Resources Planning (ERP) system, the data model only represents an abstract notion of this process. The product planning process will, however, prepare a number of business objects for elaboration during the subsequent production processes.

The first of these processes is the product engineering process, in which the content of the product is determined by the editorial staff based on the initial

[11] http://projects.ibbt.be/pisa

product specifications. After the product engineering process, the product is defined as a composition of logically and editorially constituent parts, whereby each logical unit of creative or editorial work is represented by an editorial object. In the following manufacturing engineering process, information related to the manufacturing of a product is specified and represented in manufacturing objects. Finally, during the manufacturing process, the manufacturing objects are realized in the form of audiovisual material.

The following subsections describe how the news production process is represented using this generic media production model. We introduce a number of extensions specific for news production and indicate how the corresponding information sources are mapped onto the model.

4.1 Product Engineering

An editorial object, as defined in [17], represents a unit of editorial work defined during the product engineering phase. For the news production process, we implemented three specialization subtypes of editorial object: *News*, *NewsStory*, and *NewsItem*. A news story represents a topic that is to be covered during a news broadcast. A story can consist of several news items. A news item is the atomic editorial object for news production and can, for example, correspond to an interview or a news anchor reading an introductory text from the autocue. Information contained in editorial objects is for example the story title, broadcast date, relevant keywords, etc.

Fig. 3 depicts the UML class diagram of the introduced editorial objects. The *EditorialObject* class can be reflexively associated through a many-to-many relationship. As a result, a story can be part of a news broadcast. However, a story does not need to be related to a news editorial object, as some stories are developed without eventually being part of a news broadcast. Note that application logic is responsible for prohibiting the occurrence of invalid relations, such as the fact that a news broadcast cannot be part of a news item.

The *Rundown* class is used to specify the editorial content of a news broadcast, which includes the list of news items to be broadcast, and the order in which these items are scheduled. Hence, a Rundown is attached to the News editorial object as a *EditorialObjectDescription* subclass. Similarly, anchor texts and other

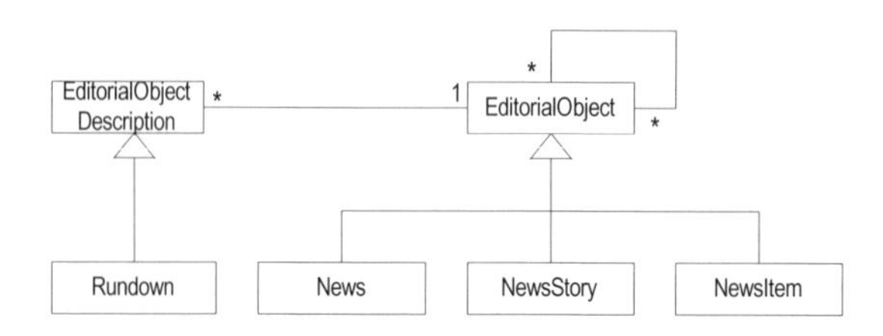

Fig. 3. The *EditorialObject* class, with subclasses *News*, *NewsStory*, and *NewsItem*, and associated *Rundown* description class

metadata that define the editorial content of the more granular news item are also incorporated into an EditorialObjectDescription. The data model supports the representation of iterative versions of EditorialObjectDescriptions to model rundowns and news items that change over time. However, in VRT's current news production process, only the final versions are effectively stored.

4.2 Manufacturing Engineering

Considering the strict deadlines and well-established format of news production, there is little time and need for extended preparations and premeditation concerning the cinematography of news items. As such, the manufacturing engineering layer of the data model is of limited use since the translation of editorial object semantics to a media object is straightforward and does not require e.g., the accurate definition of camera positions by means of storyboarding or previsualization, as is often the case with more elaborate media production processes such as drama production.

4.3 Manufacturing

Due to the lack of manufacturing engineering in current day news production, objects from the product engineering layer can be related directly to objects in the lowest manufacturing layer. In the manufacturing phase, editorial objects are materialised into (audiovisual) *Media Objects*. In news production, different versions of a media object can be generated and stored. For example, a High Definition (HD) version and a downscaled version can be realized, containing the same audiovisual content. Every version is then represented by an instance of the *MediaObjectInstance* class and related to the corresponding Media Object, as illustrated in Fig. 4.

The *Materialisation* class associates an *EditorialObject* with a *MediaObject*, and contains information specifying how the editorial object was manufactured into the media object in question. An example of information that belongs to a materialization object is the name of the camera operator. The *TimeBasedAnnotation* class provides media objects with time-based annotations. For example,

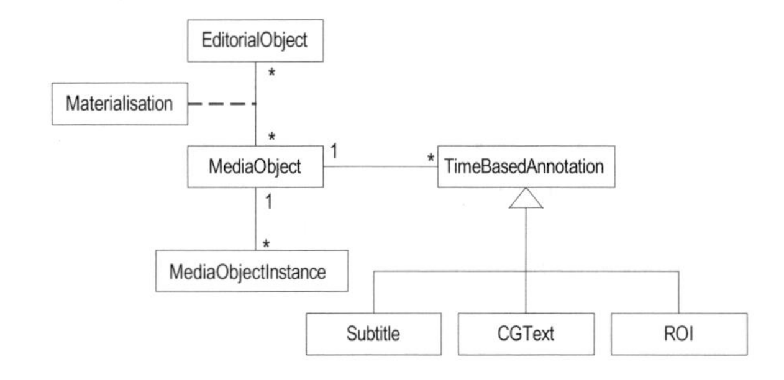

Fig. 4. The *MediaObject* class and it's relation with the *EditorialObject* class

a media object can be annotated with a subtitle or a CG text together with its associated appearance and disappearance time.

4.4 Data Source Mapping

The information from the selected data sources needs to be mapped to our proposed data model. Note that we keep the sources of the different kinds of information, since the provenance of this information is crucial for future reuse[12].

- **iNEWS:** As already mentioned, iNEWS is used as a management tool for news production. Therefore, iNEWS is used for defining instances of the *News*, *NewsStory* and *NewsItem* editorial objects. The order of the news items is defined through an instance of the *Rundown* object. Anchor text belongs to the *NewsItem* editorial object as this is generated during the product engineering process. Open subtitles and CG text are implemented as instances of the corresponding subclasses of the *TimeBasedAnnotion* class, because this information has associated timing information.
- **Ardome:** Items stored in Ardome are represented through instances of the *MediaObjectInstance* class. Metadata present in Ardome belongs to the corresponding editorial objects that are related to these instances.
- **News Agency Provided Metadata:** Information taken from the NewsML-G2 documents such as title and textual descriptions is added to the corresponding instance of the *NewsItem* editorial object.
- **Swift:** Closed subtitles are represented as instances of the *Subtitle* class.
- **Basis:** Descriptive information present in a Basis record such as title, description, and keywords is associated with the corresponding editorial object. Information related to the realisation such as the camera operator, director, and recording date is provided as part of the *Materialisation* class. Technical information such as the video coding format is provided to the corresponding instance of the *MediaObjectInstance* class. Information that is common for all versions of a materialisation is represented as part of the *MediaObject* class. An example of information present in the *MediaObject* class is rights information.

5 Enabling Semantic Search

Unifying all generated information through the use of a common data model improves the potential of a search application significantly. However, by connecting this information with external data sets, even more intelligent search operations are enabled. The following sections describe the used architecture and give an overview of how external data sets are connected. The subsequent sections then illustrate some added functionality that is obtained by following this approach.

---

[12] http://www.w3.org/2005/Incubator/prov/wiki/Presentations_on_State_of_the_Art

5.1 Architecture

The data model was formalized into an OWL ontology and all relevant information from the selected data sources discussed in Section 3 was converted into a representation according to this formalized data model and stored in a RDF triple store. The resulting RDF data present in the triple store can then be queried using the SPARQL query language. By formalizing the data model and the corresponding information from the information sources, an unambiguous and machine processable representation of the available information is obtained. This in turn enables the use of reasoners, which can derive new facts based on the provided information.

The use of Semantic Web technologies also allows us to connect with other data sources using the Linked Open Data principles [2]. This enables us to use information from external data sets in order to make more intelligent search applications. Therefore, the search application makes use of a query facade which in turn uses data from several data sets according to the Linked Open data principles. The resulting architecture is illustrated in Fig. 5.

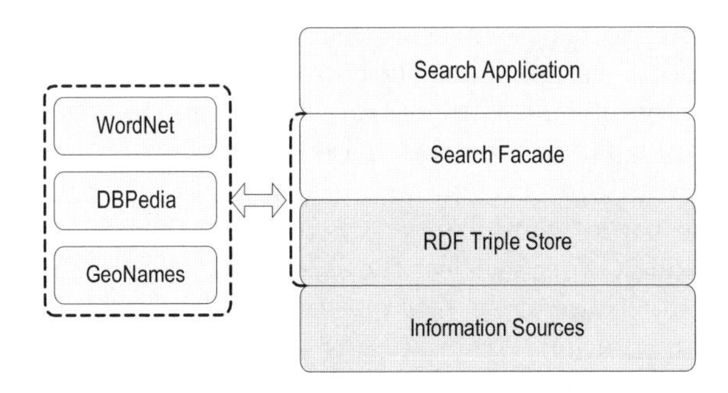

Fig. 5. Overview of the used architecture to enable semantic search

5.2 Connecting with the Linked Open Data Cloud

In order to be able to use information formalized in external data sets, concepts defined in our data model must be linked with the corresponding concepts defined in the external data sets. A valuable data set which is considered an important linking hub in the Linked Open Data cloud is DBPedia [4]. DBPedia is already used by other broadcasters [11] and proves to be a valuable data source. We use DBPedia to link concepts representing persons and other concepts, except for locations. For locations, the GeoNames[13] data set is used, providing a formalized representation of geographic locations.

As a starting point, we use the Basis keywords as interlinking mechanism with the other data sets. As already mentioned, every Basis record contains a number of relevant keywords taken from the keywords thesaurus. The keywords can directly be linked to concepts defined in other data sets. For other fields

[13] http://www.geonames.org

containing textual data such as the Basis description field and Swift subtitles, Named Entitiy Recognition (NER) [15] must first be applied. NER is part of future work that will be performed in order to extend the set of relevant entities.

We formalized the keywords thesaurus in SKOS [13]. Consequently, every concept defined in the thesaurus corresponds to a URI. Also, the relationships defined between different concepts are also present in the SKOS representation. The relations defined between concepts in the thesaurus are important in the disambiguation of concepts during mapping. For example, the concept 'apple' defined in the keywords thesaurus has the related concepts 'iMac' and 'taligent'. These related concepts are then used to select the correct corresponding concept defined in DBPedia (`http://dbpedia.org/resource/Apple_Inc.` instead of `http://dbpedia.org/resource/Apple`).

As already mentioned, locations are mapped to the corresponding concept defined in GeoNames. Again, we make extensive use of the relations defined in the thesaurus. For example, the concept 'Parijs' (Eng. Paris), has a used-for-relationship with the concept 'Paris'. Also, 'Parijs' is defined as a narrower term of 'Frankrijk' (Eng. France). With this added information, it is possible to select the corresponding concept from GeoNames.

However, for many concepts little or even no relationships at all are defined in the thesaurus. In this case, selecting the corresponding concept from an external data set can be very difficult as there is no additional context present that can be used for disambiguation. For this reason, statistics are currently collected from the Basis records, in order to get a set of keywords that most frequently co-occur with another keyword. These keywords will then be used as additional context information in order to select the correct concept from the external data set. Note also that because of the fact that the majority of keywords are defined in Dutch, often an additional translation is needed for successful mapping with concepts from external data sets, e.g. DBPedia.

By formalizing the keywords thesaurus and connecting it's concepts with external data sets, new functionalities are obtained. Some of these are illustrated in the following sections.

5.3 Suggesting Alternative Queries

The result set of a query depends on the user input. When a user enters a general keyword, the result set can be too large. In order to limit this result set, other keywords having a narrower-than relationship with the entered keyword can be displayed as a suggestion. On the other hand, when a query results in an empty result set, an alternative keyword can be suggested which has results.

5.4 Lexical Information

As the thesaurus is maintained by hand, it is difficult to include all possible information related to an introduced keyword. Consequently, related information such as abbreviations and synonyms are often not included in the thesaurus. However, this makes it more difficult for a user to find results, as they have to know which words are included in the thesaurus. In order to provide a better search experience, a connection was made with a formalized version of the WordNet lexical database[14]. Note that this data set is also linked with Cornetto[15], a lexical semantic database for the Dutch language.

When a user enters a keyword, additional information can then be retrieved and included in the query in order to optimize the search operation. For example, when a user enters 'populatie' (Eng. population) the result set of this query would be small as this concept is not included in the thesaurus. However, it can be found in Cornetto that the concept 'populatie' has similar meaning to 'bevolking' (Eng. inhabitants), which is included in the thesaurus.

5.5 User Query Evaluation

When a user enters keywords, reasoning can be performed in order to try to find out what a user really searches for. For example, when a user enters the keywords 'vice president Barack Obama', it can be derived that the user possibly searches media related to Joe Biden but maybe he does not recall his name. Although Joe Biden is present as a concept in the thesaurus, no relations are present indicating that Joe Biden is the vice president for Barack Obama.

In order to be able to show results having Joe Biden as a keyword, we use information available in external data sets as follows. When a query with multiple keywords is performed, we try to find a subject (or object) for which a property and a related object (or subject) exists corresponding with the entered keywords as follows.

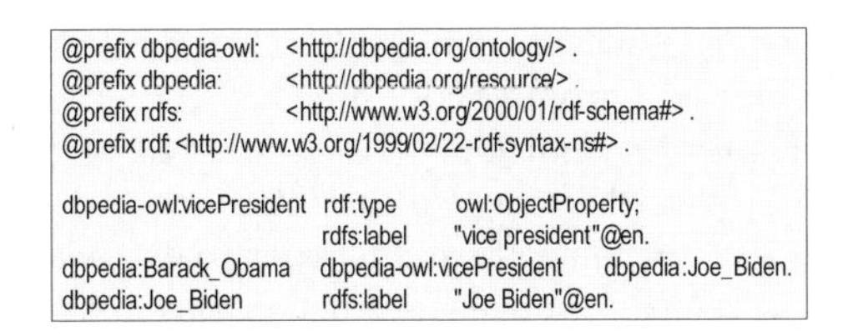

Fig. 6. Selected triples from DBPedia used to optimize the result set related to the example query

From the entered keywords we retrieve the words Barack Obama and search for a corresponding concept in the formalized thesaurus having this as a label. This concept is present in the thesaurus and is already linked with the corresponding concept in DBPedia, thus allowing the use of that information available from DBPedia. In the following step, we try to map the other entered keywords with a property occurring in a triple having the concept of Barack Obama either as its subject or object. Fig. 6 illustrates that the property

[14] http://semanticweb.cs.vu.nl/lod/wn30/
[15] http://www2.let.vu.nl/oz/cltl/cornetto/

dbpedia-owl:vicePresident has as label vice president, allowing the identification of the needed property. Then we search for triples with the selected property and where Barack Obama appears either as subject or object. Finally, we obtain the concept dbpedia:Joe_Biden, which is also linked with the corresponding concept from the keywords thesaurus. This enables the inclusion of media provided with the keyword Joe Biden, as the user ultimately wanted.

6 Conclusions and Future Work

In this paper, we proposed an architecture to optimize media search in a news production environment. Therefore, we evaluated VRT's news production process and identified data sources that contain information which can be used during media search. A data model was developed in order to uniformly represent the information contained in the data sources. The data model was implemented as an OWL ontology, enabling an unambiguous and machine processable representation of the information. This information is stored into a RDF triple store and queried by search applications using the SPARQL query language. Using Semantic Web technologies, a connection could be made with other valuable data sets that follow the Linked Data principles, e.g. DBPedia and GeoNames. Through the connection of all these information sources, advanced search functionalities are enabled and semantic search is obtained.

Future work includes an evaluation of the performance and efficiency of the developed search applications. Also, other valuable data sets from the Semantic Web will be evaluated. Relevant data sets will then be connected with the data model in order to further improve the media search. Currently, we are also researching how time-related concepts can be effectively represented and how new, inferred information can be automatically inserted and stored according to the data model. Named Entity Recognition (NER) tools will be evaluated in order to extract more relevant keywords from textual data such as subtitles. We are also implementing speech recognition software in order to extract timing information (e.g. when a certain keyword is spoken). Other feature extraction tools such as face recognition software will also be evaluated.

Acknowledgements. The research activities as described in this paper were funded by Ghent University, VRT-medialab, the Interdisciplinary Institute for Broadband Technology (IBBT), the Institute for the Promotion of Innovation by Science and Technology in Flanders (IWT), the Fund for Scientific Research Flanders (FWO-Flanders), and the European Union.

References

1. British Broadcasting Corporation (BBC). Standard Media Exchange Framework (SMEF) Data Model, v1.10 (2005), `http://www.bbc.co.uk/guidelines/smef/`
2. Berners-Lee, T.: Design Issues: Linked Data (2006), `http://www.w3.org/DesignIssues/LinkedData.html`

3. Bizer, C., Heath, T., Berners-Lee, T.: Linked Data - The Story So Far. International Journal on Semantic Web and Information Systems (2009)
4. Bizer, C., Lehmann, J., Kobilarov, G., Auer, S., Becker, C., Cyganiak, R., Hellmann, S.: Dbpedia - a crystallization point for the web of data. Web Semantics: Science, Services and Agents on the World Wide Web 7(3), 154–165 (2009)
5. Brown, M.G., Foote, J.T., Jones, G.J.F., Jones, K.S., Young, S.J.: Automatic content-based retrieval of broadcast news. In: Proceedings of ACM Multimedia 1995, San Fransisco, CA, USA, pp. 35–43 (November 1995)
6. de Jong, F.M.G., Westerveld, T., de Vries, A.P.: Multimedia search without visual analysis: the value of linguistic and contextual information. IEEE Transactions on Circuits and Systems for Video Technology 17(3), 365–371 (2007)
7. EBU. P/Meta Metadata Exchange Scheme v1.1. Technical Report 3295 (June 2005), http://www.ebu.ch/en/technical/metadata/specifications/
8. Gauvain, J.-L., Lamel, L., Adda, G.: Transcribing broadcast news for audio and video indexing. Communications of the ACM 43(2), 64–70 (2000)
9. Hargreaves, I., Thomas, J.: New news, old news. ITC and BSC research publication (October 2002)
10. International Press Telecommunications Council. NewsML-G2 v.2.2 (November 2008), http://www.iptc.org
11. Kobilarov, G., Scott, T., Raimond, Y., Oliver, S., Sizemore, C., Smethurst, M., Bizer, C., Lee, R.: Media Meets Semantic Web – How the BBC Uses DBpedia and Linked Data to Make Connections. In: Aroyo, L., Traverso, P., Ciravegna, F., Cimiano, P., Heath, T., Hyvönen, E., Mizoguchi, R., Oren, E., Sabou, M., Simperl, E. (eds.) ESWC 2009. LNCS, vol. 5554, pp. 723–737. Springer, Heidelberg (2009)
12. Mannens, E., Verwaest, M., Van de Walle, R.: Production and multi-channel distribution of news. Multimedia Systems (14), 359–368 (2008)
13. Miles, A., Matthews, B., Wilson, M., Brickley, D.: SKOS core: Simple Knowledge Organisation for the Web. In: Proceedings of the 2005 International Conference on Dublin Core and Metadata Applications, pp. 1–9. Dublin Core Metadata Initiative, Madrid (2005)
14. Netter, K., de Jong, F.: Olive: Speech based video retrieval. In: Proceedings of CBMI 1999, Toulouse, France, pp. 75–80 (October 1999)
15. Nguyen, H.T., Cao, T.H.: Named entity disambiguation: A hybrid statistical and rule-based incremental approach. In: Domingue, J., Anutariya, C. (eds.) ASWC 2008. LNCS, vol. 5367, pp. 420–433. Springer, Heidelberg (2008)
16. Smith, J.R., Schirling, P.: Metadata Standards Roundup. IEEE Multimedia 13(2), 84–88 (2006)
17. Van Rijsselbergen, D., Verwaest, M., Van De Keer, B., Van de Walle, R.: Introducing the Data Model for a Centralized Drama Production System. In: Proceedings of the IEEE Intl. Conference on Multimedia & Expo 2007, pp. 615–618 (July 2007)

Integration of Existing Multimedia Metadata Formats and Metadata Standards in the M3O

Daniel Eißing, Ansgar Scherp, and Carsten Saathoff

WeST, University of Koblenz-Landau, Germany
{eissing,scherp,saathoff}@uni-koblenz.de
http://west.uni-koblenz.de

Abstract. With the Multimedia Metadata Ontology (M3O), we have developed a sophisticated model for representing among others the annotation, decomposition, and provenance of multimedia metadata. The goal of the M3O is to integrate existing metadata standards and metadata formats rather than replacing them. To this end, the M3O provides a scaffold needed to represent multimedia metadata. Being an abstract model for multimedia metadata, it is not straightforward how to use and specialize the M3O for concrete application requirements and existing metadata formats and metadata standards.

In this paper, we present a step-by-step alignment method describing how to integrate and leverage existing multimedia metadata standards and metadata formats in the M3O in order to use them in a concrete application. We demonstrate our approach by integrating three existing metadata models: the Core Ontology on Multimedia (COMM), which is a formalization of the multimedia metadata standard MPEG-7, the Ontology for Media Resource of the W3C, and the widely known industry standard EXIF for image metadata.

1 Introduction

A densely populated jungle with a myriad of partially competing species of different colors and size—this might be a good characterization of today's world of multimedia metadata formats and metadata standards. Looking at the existing metadata models like [1,2,3,4,5] and metadata standards such as [6,7,8,9,10], we find it hard to decide which of them to use in a complex multimedia application. They focus on different media types, are very generic or designed for a specific application domain, and overlap in the functionality provided.

However, building a complex multimedia application often requires using several of these standards *together*, e.g., when different tools have to be integrated along the media production process [11]. The integration among tools requires interoperability of different metadata standards, which is a requirement that is not sufficiently satisfied by existing formats and standards. With XMP [7], there exists an important initiative to enable interoperability along the production process of images. Nevertheless, this work is limited with respect to the functionality provided and focuses on the media type image only [12]. Overall, the

T. Declerck et al. (Eds.): SAMT 2010, LNCS 6725, pp. 48–63, 2011.

XMP initiative is an important step but more is required to facilitate multimedia metadata interoperability along the media production process.

To solve this problem, we have developed the Multimedia Metadata Ontology (M3O) [12]. The M3O is a sophisticated model for representing among others the annotation, decomposition, and provenance of multimedia content and multimedia metadata. The goal of the M3O is to provide a framework for the integration of existing metadata formats and metadata standards rather than replacing them. The M3O bases on a foundational ontology and by this inherits its rich axiomatization. It follows a pattern-based ontology design approach, which allows the M3O to arrange the different functionalities for representing multimedia metadata into smaller, modular, and reusable units.

However, the M3O was designed as an abstract model providing a scaffold for representing arbitrary multimedia metadata. As such, the integration of existing standards is not straightforward, and we are confronted with a gap between the formal model and its application in concrete domains. In this paper, we fill this gap and present a step-by-step alignment method describing how to integrate existing formats and standards for multimedia metadata and the M3O. We describe the tasks that have to be performed for this integration and apply the integrated ontology to a concrete modeling task. We demonstrate this integration at the example of the Core Ontology on Multimedia (COMM) [2], which is a formalization of the multimedia metadata standard MPEG-7 [9], the recently released Ontology for Media Resource [13] of the W3C, and the widely known and adopted industry standard EXIF [6] for image metadata.

2 Introduction to the Multimedia Metadata Ontology

The Multimedia Metadata Ontology (M3O) [12] provides a generic modeling framework to integrate existing multimedia metadata formats and metadata standards. The M3O is modeled as a highly axiomatized core ontology basing on the foundational ontology DOLCE+DnS Ultralight (DUL) [14]. DUL provides a philosophically grounded conceptualization of the most generic concepts such as objects, events, and information. The axiomatization is formulated in Description Logics [15].

The M3O follows a pattern-based approach to ontology design. Each pattern is focused on modeling a specific and clearly identified aspect of the domain. From an analysis of existing multimedia metadata formats and metadata standards [12], we have identified six patterns required to express the metadata for multimedia content. These patterns model the basic structural elements of existing metadata models and are the Decomposition Pattern, Annotation Pattern, Information Realization Pattern, Data Value Pattern for representing complex values, Collection Pattern, and Provenance Pattern. Basing a model like the M3O on ontology design patterns ensures a high degree of modularity and extensibility, while at the same time a high degree of axiomatization and thus semantic precision is retained. In order to realize a specific multimedia metadata format or metadata standard in M3O, these patterns need to be specialized. In the following, we discuss three patterns of the M3O in more detail, namely the *Information Realization Pattern*, *Annotation Pattern*, and *Data Value Pattern*, which we will mainly refer to in the upcoming sections.

Information Realization. The information realization pattern in Figure 1a models the distinction between information objects and information realizations [14]. Consider a digital image that is stored on the hard disk in several formats and resolutions. An information object represents the image as an abstract *concept* or *idea*, namely the information object of an image. Different files may *realize* this same abstract idea. As shown in Figure 1a, the pattern consists of the InformationRealization that is connected to the InformationObject by the realizes relation. Both are subconcepts of InformationEntity, which allows treating information in a general sense as we will see in the Annotation Pattern.

Annotation Pattern. Annotations are understood in the M3O as the attachment of metadata to an information entity. Metadata comes in various forms such as low-level descriptors obtained by automatic methods, non-visual information covering authorship or technical details, or semantic annotation aiming at a formal and machine-understandable representation of the contents. Our Annotation Pattern models the basic structure that underlies all types of annotation. This allows for assigning arbitrary annotations to information entities while providing the means for modeling provenance and context. In Figure 1b, we see that an annotation is not modeled as a direct relationship between some media item and an annotation. It is defined by a more complex structure, which is inherited by the Descriptions and Situations Pattern of DUL. Basically, a Descriptions and Situations Pattern is two-layered. The *Description* defines the structure, in this case of an annotation, which contains some entity that is annotated and some entity that represents the metadata. The *Situation* contains the concrete entities for which we want to express the annotation. The pattern allows us to add further concepts and entities into the *context* of an annotation, e.g.,

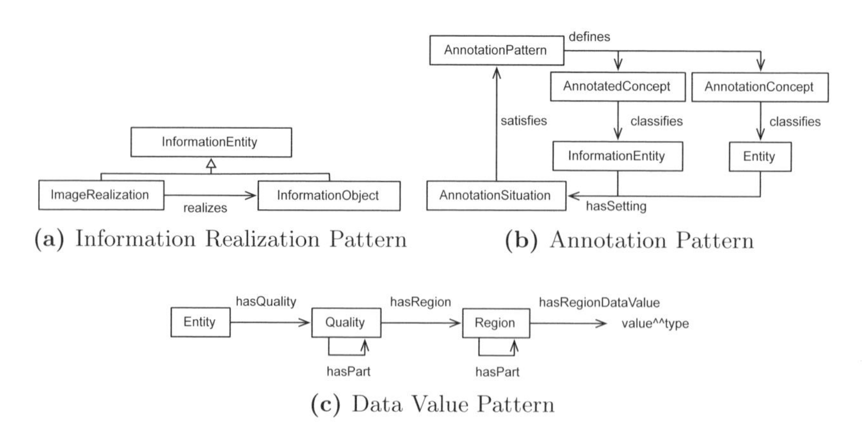

(a) Information Realization Pattern

(b) Annotation Pattern

(c) Data Value Pattern

Fig. 1. Three Example Patterns of the Multimedia Metadata Ontology (M3O)

expressing provenance or confidence information. On the top half, we see that the AnnotationPattern defines an AnnotatedConcept and an AnnotationConcept. The AnnotatedConcept classifies an InformationEntity and thus expresses that the information entity is the subject of the annotation. The AnnotationConcept classifies some Entity, which identifies this entity as the annotation or metadata. The entity can be some complex data value, e.g., representing some low-level features represented using the Data Value Pattern, but also some concept located in a domain ontology such as DBpedia[1]. All the entities have as setting the AnnotationSituation, which satisfies the AnnotationPattern.

Data Value Pattern. In ontologies we mainly use abstract concepts and clearly identifiable individuals to represent data. However, we also need the means to represent concrete data values such as strings and numerical values. In DUL, there exists the concept Quality in order to represent attributes of an Entity, i.e., attributes that only exist together with the Entity. Regions are used to represent the values of an Quality and the data space they come from. The Data Value Pattern depicted in Figure 1c assigns a concrete data value to an attribute of that entity. The attribute is represented by the concept Quality and is connected to the Entity by the hasQuality property. The Quality is connected to a Region by the hasRegion relation. The Region models the data space the value comes from. We attach the concrete value to the Region using the relation hasRegionDataValue. The data value is encoded using typed literals, i.e., the datatype can be specified using XML Schema Datatypes [16].

3 Alignment Method

This section introduces our method for aligning multimedia standards and multimedia formats with the M3O. The method has been derived from our experiences applying and specializing the M3O for three existing multimedia formats and standards, namely COMM [2], the Ontology for Media Resource [13], and EXIF [6].

In contrast to automatic, adaptive, or machine learning approaches for ontology alignment [17,18,19], we conduct a pure manual alignment, as only a manual alignment ensures the high quality of the integration and minimizes ambiguities and imprecise matching. We consider the time and effort for manual alignment manageable, as we assume that each metadata format or standard has to be aligned only once and that updates to the integrated formats or standards will be infrequent and mostly incremental.

For the alignment, we propose an iterative four-step alignment method that helps ontology engineers to integrate existing metadata standards and metadata formats. In each iteration, we consecutively evolve the alignment of the format or standard with the M3O. Following an iterative approach, we are able to identify, analyze, and flexibly react to problems and challenges encountered during previous iterations of the alignment.

[1] http://dbpedia.org/

Each iteration consists of four steps. The first step targets the understanding of the format or standard to be integrated. The second step reorders concepts into coherent groups. The third step maps concepts and structure with the M3O. The fourth step proves and documents the validity of the alignment, and finalizes the iteration.

To introduce our alignment method, we proceed as follows: For each step, we first outline the goals and provide a brief summarization. Subsequently, we describe the core tasks to be performed within the step, and provide concrete examples that show its relevance and application for concrete ontologies.

3.1 Step 1: Understanding

Summary. A precise understanding of the metadata format or standard to be integrated is an import prerequisite for aligning it with the M3O. Consequently, the first step of alignment is an in-depth analysis of the structure and core concepts of the model at hand. While this advise may seem obvious, this is a task easily underestimated and problems neglected at an early stage can cause time-consuming problems along the integration process. Additional documentation, if available, will help to (re-)produce the overall structure not explicitly expressed in the formal specification.

Detailed Description and Examples. In general, we have found three distinct modeling approaches to be very common for multimedia metadata formats and metadata standards.

Pattern-based. Pattern-based ontologies, e.g., COMM, provide a high degree of axiomatization and structure in a formal and precise way. Through our analysis, we understand the patterns used and the functionality they provide. This allows us to compare the patterns of the ontology to be integrated with those provided by the M3O.

Predicate-centric. In a predicate-centric approach, as followed e.g. by the Ontology for Media Resource, the ontology is mainly specified through a set of properties. Such a model offers very little structure in a machine readable format, e.g., in terms of conceptual relations between properties. However, by analyzing the documentation, we infer additional information about intended groupings of properties and the structural composition of the format or standard to be integrated.

Legacy Models. Other formats and standards have not yet been semantified at all. By analyzing the concepts and relations expressed in the specification of the format, we decide how the core concepts can be expressed in a formal and precise way using the M3O.

Ambiguities that are found during the initial analysis are discussed at this point. It is not our intention to revise all modeling decisions made for the ontology to be integrated. However, we consider the alignment a good opportunity to correct some of the *bad smells* [20] discovered. Once we have reached a sufficient understanding of the format or standard to be integrated, we proceed with the grouping step.

3.2 Step 2: Grouping

Summary. Ontologies should provide structural information on the relation and groupings of concepts it defines. However, although many formats or standards provide this information in their documentation, the information is sometimes lost when the models are transformed to an ontology. By using the original specifications and documentations, we are able to preserve and recreate this information grouping, and provide them through formal specification in the aligned ontology.

Detailed Description and Examples. In principle, we distinguish three forms of available grouping information:

Explicit Grouping. Pattern-based models provide an explicit grouping of concepts into coherent patterns, often accompanied by a rich axiomatization on how they relate. As an example, the definition of a color histogram annotation in COMM specifies a ColorQuantizationComponentDescriptorParameter that groups the concepts ColorComponent and NumberOfBinsPerComponent.

Implicit Grouping. For other metadata models grouping information may not be explicitly represented. This is often the case with predicate-centric approaches, e.g., the Ontology on Media Resource. In these cases, we refer to the textual documentation in order to (re-)construct the implicit groupings of the properties or classes. As an example, the documentation of the Ontology on Media Resource offers a textual description on the grouping of its properties, e.g., in terms of identification or creation. However, this information is not accessible in the RDF representation as proposed by the W3C. By defining the appropriate axioms, we have appended the implicit grouping information in a formal and explicit way, e.g., by stating that an IdentificationAnnotation hasPart some TitleAnnotation, LanguageAnnotation, and LocatorAnnotation.

Recovery of Groupings. In other cases grouping information is lost when transferring multimedia formats or standards to RDF. For example the EXIF metadata standard provides textual descriptions about groupings, e.g., in terms of pixel composition and geo location. However, this distinction got lost in the adaption to an RDF schema [21]. For the alignment with the M3O, we have reconstructed the grouping information and appended it to the model in a formal and explicit way.

Once we have provided all relevant grouping relations through a formal specification, we continue with the mapping step.

3.3 Step 3: Mapping

Summary. This step achieves the mapping of the ontology's concepts and structure to the scaffold provided by the M3O. The goal of this step is to create a working ontology, which, after validation, can be published or used as basis for further iterations.

Detailed Description and Examples. For the alignment we follow a sequence of the following three steps:

1. **Mapping of Concepts.** If some superclass of the concept to be aligned is present in both ontologies, direct mapping of concepts is feasible. This is mainly the case for ontologies that share the same foundation, e.g., COMM and the M3O, which both base on the DUL foundational ontology. All axioms of the aligned concepts are preserved as long as they are applicable through the M3O. If a concept is not applicable in the M3O, we align all dependent subclasses and references to the nearest matching M3O concept. As an example, the COMM DigitalData concept, which is a subclass of the DUL InformationObject, was removed during the alignment. The dangling dependencies and references have been resolved by subclassing or referencing the InformationObject instead.
2. **Structural Mapping.** For structural mapping, we consider the functionality of the pattern or structure to be mapped. If a pattern or structure offers the same or an equal functionality than a pattern of the M3O, we can replace the pattern. By adapting the M3O pattern, we are often able to express the same functionality using a more generic approach. As an example, COMM proposes the Digital Data Pattern to express data values in a digital domain. A similar functionality is provided by the M3O Data Value Pattern, which expresses data values through the generic concepts of Quality and Region. The COMM Digital Data Pattern can be considered a special case of expressing data values and therefor has been replaced using the M3O Data Value Pattern instead.

   In the same manner, we simplify the structural composition of the existing model by merging multiple concepts and patterns that offer the same or an equal functionality. As an example, COMM defines three annotation patterns. Each deals with a different aspect of multimedia annotation, although they vary only slightly in their structural composition. We have aligned those patterns by adapting the M3O Annotation Pattern. The domain specific concepts that result from the separation into three distinct patterns have been preserved by subclassing the corresponding concepts of the M3O Annotation Pattern. This simplifies the structure of the model, while also preserving the original functionality.
3. **Removing Unnecessary Concepts.** We finalize the mapping step by cleaning up unused dependencies from the ontology files. Concepts that either have no further relevance for the target context or are sufficiently covered by the M3O are removed at this point. An example, the COMM AnnotatedMediaRole offers an equal functionality as the M3O AnnotatedInformationRealizationConcept. We therefore have removed COMM's AnnoatedMediaRole and replaced any formal relation that involves the concept.

3.4 Step 4: Validation and Documentation

In each iteration of the alignment process, we need to check the consistency of the resulting ontology. This can be done by using a reasoner like Fact++[2] or Pellet[3]. Any problem encountered during the alignment can be resolved by reiterating the four steps of the alignment method. After proving the consistency of the resulting ontology, we finalize the process by documenting all major decisions and adjustments made during the alignment.

3.5 Summary

In this section, we have proposed a four-step method for aligning multimedia metadata formats and multimedia metadata standards with the M3O. In the following Sections 4-6, we demonstrate the alignment of three existing formats and standards by applying our method. They are the Core Ontology on Multimedia, the Ontology for Media Resource, and the EXIF metadata standard.

4 Example 1: Core Ontology on Multimedia (COMM)

The Core Ontology on Multimedia (COMM) [2] is a formal specification of the MPEG-7 metadata standard [9]. In contrast to other approaches to modeling MPEG-7 as an ontology COMM is not designed as a one-to-one mapping, but provides a set of patterns that cover the core and repetitive building blocks of MPEG-7. The central challenge of the alignment of COMM and M3O is understanding the patterns of COMM and mapping them to the scaffold provided by the M3O. This section describes the experiences and challenges of aligning COMM and the M3O, using the four-step alignment method proposed above.

4.1 Application of the Alignment Method

Understanding. COMM follows a pattern-based approach and builds on the DUL foundational ontology. Some of the core patterns, i.e., the Descriptions and Situations Pattern, are shared between COMM and the M3O. Others, e.g., the Digital Data Pattern, form major structural differences.

COMM defines five structural patterns, namely the Content Annotation Pattern, Media Annotation Pattern, and Semantic Annotation Pattern for media annotation, the Decomposition Pattern for media (de-)composition, and the Digital Data Pattern, which expresses annotations in a digital domain. Domain specific knowledge is separated from the core concepts and defined in separate ontologies, e.g., concepts concerning annotation of visual entities are defined in the *visual ontology*.

Some ambiguities that were found in the initial analysis have been resolved at this point. As an example, COMM specifies concepts such as NumberOfBinsPerComponent that are specialization of both Parameter and Region. While this may

[2] `http://owl.man.ac.uk/factplusplus/`

[3] `http://clarkparsia.com/pellet/`

not be syntactically incorrect, it violates the DnS pattern of DUL. In the DnS pattern, a Parameter parametrizes a Region. Thus, these two concepts should not have common sub-concepts. To solve this problem, we have removed the superclass relations to the Parameter concept and introduced a parametrizes relation. For example, COMM specified a ColorComponent and NumberOfBinsPerComponent, which are subclasses of both the ColorQuantizationComponentDescriptorParameter and the Region concept. We have removed the superclass relation from the ColorComponent and NumberOfBinsPerComponent to the ColorQuantizationComponentDescriptorParameter, which instead now parametrizes these concepts.

Grouping. Following a pattern-based design, COMM already provides a rich degree of conceptual groupings and their axiomatization in a machine readable format. However, reusability can be improved by redistributing concepts among the six ontologies of COMM, *core*, *datatype*, *localization*, *media*, *visual*, and *textual* respectively. As an example, the concept RootSegmentRole, located in the COMM *core ontology*, is not used in any pattern definition and has therefore been relocated to the *localization ontology*.

Mapping. The main challenge of aligning COMM and the M3O concern the differences of the patterns used and how to relate them. Although some principles are shared between the ontologies, there are also major differences, e.g., the Digital Data Pattern of COMM and the Information Realization Pattern of the M3O.

Often COMM patterns have been replaced using a more generic pattern of the M3O. As an example, Figure 2 displays the adaptation of the COMM Digital Data Pattern through the M3O. For the alignment, we have decided that the functionality of the Digital Data Pattern, i.e., expressing data values, can be maintained by adopting the M3O Data Value Pattern instead. All related concepts have either been removed or mapped to the next matching M3O concept. As an example, the StructuredDataDescription concept has been removed as it held no further relevance in the context of the Data Value Pattern. The StructuredDataParameter concept on the other hand has been preserved as specialization of the M3O Annotation Pattern. To accommodate StructuredDataParameters with the M3O, we

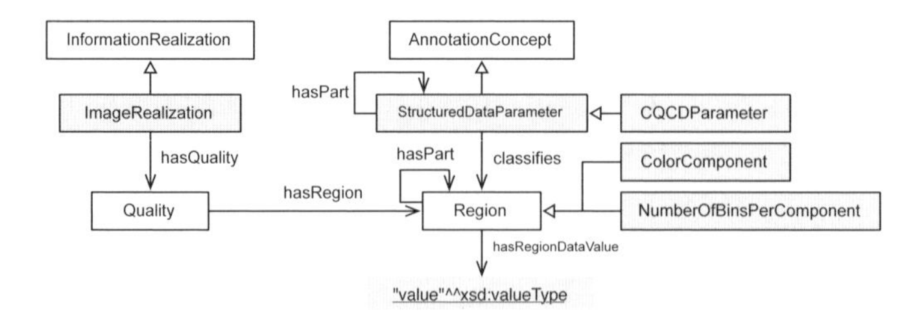

Fig. 2. Excerpt of COMM subsequently to the integration with the M3O. White boxes show the concepts of the M3O or DUL, whereas grey boxes represent concepts of COMM aligned to the M3O.

consider StructuredDataParameters as subclass of the AnnotationConcept. Through parametrizing the appropriate Region, we can constrain the range applicable for a specific StructuredDataParameter. The value itself is expressed using the hasRegionDataValue relation. In a similar manner, the three annotation patterns of COMM have been replaced through the M3O Annotation Pattern, and all dependent concepts have been mapped to the M3O Annotation Pattern instead.

Validation and Documentation. The alignment of COMM and the M3O has been validated using Fact++ and Pellet reasoner. The results have been documented in a publicly accessible wiki page available at: http://semantic-multimedia.org/index.php/COMM_integration.

4.2 Application of the Aligned Ontology

Figure 3 demonstrates the application of StructuredDataParameters using COMM aligned with the M3O. We specify a ColorQuantizationComponentDescriptorParameter (CQCDParameter) as part of the RGBHistogramAnnotationConcept. The CQCDParameter parametrizes the ColorComponents and NumberOfBinsPerComponent, which are considered part of the RGBHistogramRegion. The hasRegionDataValue relation expresses the primitive value for this annotation, e.g., an unsigned int for the NumberOfBinsPerComponent concept. Staying in line with the specification of the M3O Data Value Pattern, we consider the use of StructuredDataParameters optional. Thus, we do not specify that an AnnotationConcept must specify any StructuredDataParameters in a hasPart relation but recommend using them as they add an additional layer of formal expressiveness.

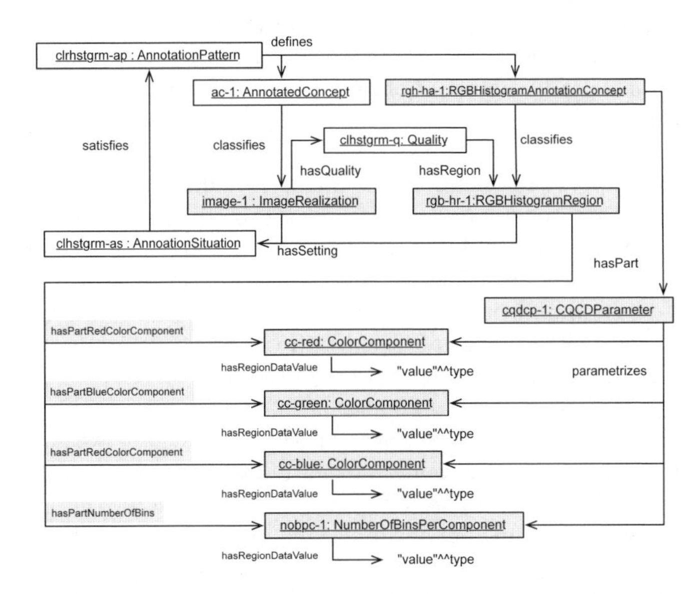

Fig. 3. Application of COMM after its integration with the M3O

5 Example 2: Ontology for Media Resource

The Ontology for Media Resource [13] developed by the W3C defines a core vocabulary for multimedia annotation. The ontology targets at an unifying mapping of common media formats like EXIF [6] or Dublin Core [10]. The core challenge for this alignment of the Ontology for Media Resource concerns the mapping of properties to either information object or information realization as provided by the Information Realization Pattern of the M3O.

5.1 Application of the Alignment Method

Understanding. The Ontology for Media Resource presents a property-centric approach to ontology modeling and consists of 28 predicates including properties like *title* and *language*. Some properties are specified in further detail, e.g., through *role* or *type* properties. Only entities such as multimedia items and persons are represented as resources. Any other information such as roles or types are represented using primitive values, e.g., strings. The Ontology for Media Resource defines neither structural patterns nor formal logical constraints beyond the domain and range specification for each property. Unlike the M3O, there is no distinction between information object and information realization.

Grouping. The Ontology for Media Resource's documentation on the web provides a number of conceptual groupings for certain aspects of multimedia description, e.g., identification or fragmentation. However, this information is not accessible in a machine readable format. With the alignment of the Ontology for Media Resource and the M3O, we have provided grouping information by defining the appropriate axioms. For example an IdentificationAnnotation concept hasPart some TitleAnnotationConcept, LanguageAnnotationConcept, and LocatorAnnotationConcept.

Mapping. For mapping the ontology for Media Resource to the M3O, we define a subconcept of the AnnotationConcept for each predicate of the ontology. For example, we define a LocatorAnnotationConcept to match the *locator* property. Concrete values are expressed using the Data Value Pattern of the M3O. To this end, we define appropriate Region concepts. In the case of the LocatorAnnotationConcept, we define a LocatorRegion with the property hasRegionDataValue and an URI specifying a concrete location on the web.

Of primary concern for this alignment is the mapping with the Information Realization Pattern. By taking into account the difference between information objects and information realizations, we can improve semantic precision of the aligned ontology. To this end, we examine each attribute of the Ontology for Media Resource for its inherent meaning and constrain it to the appropriate concept of the Information Realization Pattern of the M3O. As an example, the *locator* property of the Ontology for Media Resource annotates media files that are locatable on the web. This is a quality only applicable for information realizations and is expressed in the definition of the LocatorAnnotationConcept.

We express the *type* property of the Ontology for Media Resource through specialization, e.g., by specifying an ImageRealization, a subclass of the InformationRealization, as the type for the considered media item. Finally the *fragments* facet of the Ontology for Media Resource has been modeled using the Decomposition Pattern of the M3O. The functionality indicated by the *namedFragments* property can be obtained by decomposing multimedia items using the M3O Decomposition Pattern and by using the M3O Annotation Pattern to annotate the resulting fragment with a FragmentLabelAnnotationConcept.

Consistency Checking and Documentation. The resulting ontology has been validated using Fact++ and Pellet.

5.2 Application of the Aligned Ontology

Figure 4 demonstrates the application of the aligned ontology. We explicitly distinguish between an ImageObject and an ImageRealization that realizes the ImageObject. The specific type for each media is expressed through specialization of the corresponding InformationObject and InformationRealization concepts. The ImageObject is annotated with some TitleAnnotationConcept, where the title "Mona Lisa" is expressed using the Data Value Pattern. The ImageRealization is annotated with some LocatorAnnotationConcept that parametrizes a Region for an URI locatable on the web.

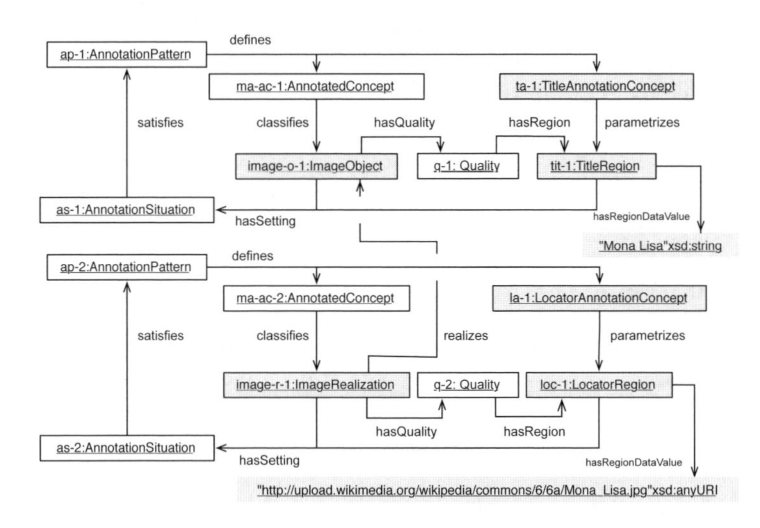

Fig. 4. Application of the Ontology on Media Resource after the alignment

6 Example 3: EXIF

EXIF is a common metadata standard for images and supports mainly technical metadata [6]. It is embedded directly into media assets such as JPEG files. The following section presents the alignment of EXIF and the M3O.

6.1 Application of the Alignment Method

Understanding. The key-value based metadata specified in EXIF is binary encoded into the header of, e.g., JPEG files. Consequently the mapping of the non-semantified concepts onto the scaffold of the M3O posed the major challenge for this particular alignment. Thus, for aligning EXIF and the M3O, we first needed to semantify the key-value pairs of EXIF.

Grouping. The EXIF metadata standard has been translated to a RDF Schema [21] by the W3C through an one-to-one mapping. Here, each key of the EXIF specification has been directly mapped to a corresponding property. This approach ignores the groupings of metadata keys that is provided in the original EXIF specification such as pixel composition and geo location. For the alignment, we have reconstructed this grouping information.

Mapping. Mapping EXIF to the M3O follows a similar procedure as conducted with the mapping of the Ontology of Media Resource. Special consideration is provided on how to map EXIF properties to information objects and information realizations. For example, locations have been constrained to information objects, as they convey information on where the original picture has been taken. Image resolutions describe a quality of a concrete realization, e.g., a JPEG file, and are therefore associated with the information realization instead. Specific properties can be referred to by using preexisting vocabularies, e.g., the WGS84 vocabulary [22] for GPS information. As EXIF restrains itself to describing qualities of multimedia items all keys have been mapped as specialization of the M3O Annotation Pattern.

Consistency Checking and Documentation. We have tested the validity of the resulting ontology using Fact++ and Pellet.

6.2 Application of the Aligned Ontology

The example as shown in Figure 5 defines an EXIFAnnotationPattern concept that allows us to represent EXIF compliant annotations. In this case, the

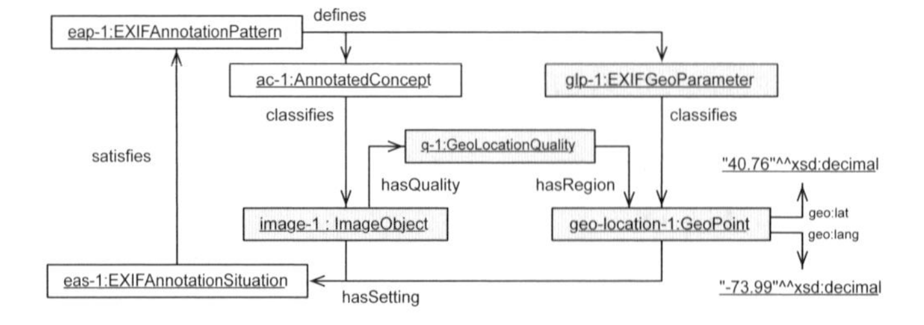

Fig. 5. Application of the EXIF metadata standard using M3O

EXIFAnnotationPattern defines an EXIFGeoParameter that parametrizes a GeoPoint. Within this construct, we accumulate all parameters that can be specified in regards to *GPS Attribute Information* as specified in EXIF. Going conform with the Data Value Pattern, we express the GeoPoint through geo:lat and geo:long, which specify primitive values of type xsd:decimal. In this case, we want to represent the location at which the image was created and thus attach the information to the information object.

7 Related Work

We shortly review the state of aligning different metadata standards. For a detailed analysis of existing metadata formats and metadata standards, we refer to the original M3O paper [12] and specifically to the report of the W3C Multimedia Semantics Incubator Group [23] and the overview of the current W3C Media Annotations Working Group [24].

Numerous metadata models and metadata standards with different goals and backgrounds have been proposed in research and industry. Most focus on a single media type such as image, text, or video, differ in the complexity of the data structures they provide, and provide partly overlapping and partly complementary functionality. With standards like EXIF [6], XMP [7], and IPTC [8] we find metadata models that provide (typed) key-value pairs to represent metadata of the media type image. An example for a more complex standard is MPEG-7 [9]. MPEG-7 provides a rich set of complex descriptors that mainly focus on expressing low-level features of images, audio, and video.

The drawbacks of these standards are the lacking interoperability and the alignment between them. Harmonization efforts like the Metadata Working Group [25] or the Media Annotations Working Group [24] try to tackle these issues and develop a common vocabulary. However, they remain on the same technological level and do not extend their effort beyond the single media type of images and do not provide a generic framework suitable for arbitrary metadata formats and arbitrary metadata standards. XMP aims at an integrated standard for image metadata. However, it tackles the problem from a different point of view. While XMP also aims at providing a framework for multimedia metadata, it focusses on images only and does not consider other media types or structured multimedia content. Another major difference is that XMP stays on the level of standards such as EXIF or IPTC and does not take into account requirements such as provenance of metadata, decomposition, or information realization.

Several approaches have been published providing a formalization of MPEG-7 as an ontology [26], e.g., by Hunter [1] or the Core Ontology on Multimedia [2]. Although these ontologies provide clear semantics for the multimedia annotations, they still focus on MPEG-7 as the underlying metadata standard. More importantly, these ontologies basically provide a formalization of MPEG-7, but do not provide for the integration of different standards.

The alignment method discussed in this paper is fully manual. There are numerous publications about (semi-)automatic alignment and matching methods [17, 18, 19]. However, these methods typically do not provide the high accuracy we require in an alignment of different metadata standards and are usually applied to problems such as ontology learning or the alignment of domain models. The M3O is a core ontology, i.e., an ontology providing some underlying structure for specific aspects of an application. The method presented in this paper shows how to align existing metadata formats and metadata standards with such a core ontology. The goal of this work is producing a specialization of the M3O that inherits the same level of formal precision and conciseness. Achieving this goal with an automatic method currently seems not realistic.

8 Conclusions

In this paper, we have shown how the generic scaffold provided by the Multimedia Metadata Ontology (M3O) can be specialized to integrate existing multimedia metadata formats and metadata standards. To this end, we have developed a four-step alignment method that describes the tasks to be performed. We have demonstrated the applicability of our approach at the example of three existing metadata models, the Core Ontology on Multimedia, the Ontology for Media Resource of the W3C, and the industry standard EXIF for image metadata. The experiences made in conducting the alignment with the M3O have been described. The results are also continuously documented on our wiki: http://www.semantic-multimedia.org/index.php/M3O:Main#Mappings

Acknowledgment. This research has been co-funded by the EU in FP7 in the WeKnowIt project (215453).

References

1. Hunter, J.: Enhancing the semantic interoperability of multimedia through a core ontology. IEEE Transactions on Circuits and Systems for Video Technology 13(1), 49–58 (2003)
2. Arndt, R., Troncy, R., Staab, S., Hardman, L., Vacura, M.: COMM: designing a well-founded multimedia ontology for the web. In: Aberer, K., Choi, K.-S., Noy, N., Allemang, D., Lee, K.-I., Nixon, L.J.B., Golbeck, J., Mika, P., Maynard, D., Mizoguchi, R., Schreiber, G., Cudré-Mauroux, P. (eds.) ASWC 2007 and ISWC 2007. LNCS, vol. 4825, pp. 30–43. Springer, Heidelberg (2007)
3. Markkula, M., Sormunen, E.: End-user searching challenges indexing practices in the digital newspaper photo archive. Information Retrieval 1(4) (2000)
4. Hollink, L., Schreiber, A.T., Wielinga, B.J., Worring, M.: Classification of user image descriptions. International J. of Human-Computer Studies 61(5), 601–626 (2004)
5. Hollink, L., Schreiber, G., Wielinga, B.: Patterns of semantic relations to improve image content search. Web Semantics: Science, Services and Agents on the World Wide Web 5(3) (2007)
6. Japan Electronics and Information Technology Industries Association: Exchangeable image file format for digital still cameras: Exif version 2.2 (2002)

7. Adobe Systems, Inc.: XMP Specifications (2008), http://www.adobe.com/devnet/xmp/
8. International Press Telecommunications Council: "IPTC Core" Schema for XMP Version 1.0 Specification document (2005), http://www.iptc.org/
9. MPEG-7: Multimedia content description interface. Technical report, Standard No. ISO/IEC n15938 (2001)
10. Dublin Core Metadata Initiative: DCMI Metadata Terms (2008), http://dublincore.org/documents/dcmi-terms/
11. Hardman, L., Obrenovic, Z., Nack, F., Kerhervé, B., Piersol, K.W.: Canonical processes of semantically annotated media production, vol. 14(6), pp. 327–340 (2008)
12. Saathoff, C., Scherp, A.: Unlocking the semantics of multimedia presentations in the web with the multimedia metadata ontology. In: WWW 2010, pp. 831–840. ACM, New York (2010)
13. W3C: Ontology for media resource 1.0 (2010), http://www.w3.org/TR/mediaont-10/
14. Borgo, S., Masolo, C.: Foundational choices in DOLCE. In: Handbook on Ontologies, 2nd edn. Springer, Heidelberg (2009)
15. Baader, F., Calvanese, D., McGuinness, D.L., Nardi, D., Patel-Schneider, P.F. (eds.): The Description Logic Handbook. Cambridge University Press, Cambridge (2003)
16. Biron, P.V., Malhotra, A.: XML Schema Part 2: Datatypes Second Edition, W3C Recommendation (2004), http://www.w3.org/TR/xmlschema-2/
17. Euzenat, J., Shvaiko, P.: Ontology matching. Springer, Heidelberg (2007)
18. Ehrig, M.: Ontology Alignment: Bridging the Semantic Gap. Semantic Web and Beyond, vol. 4. Springer, Berlin (2007)
19. Blomqvist, E.: Ontocase-automatic ontology enrichment based on ontology design patterns. In: Bernstein, A., Karger, D.R., Heath, T., Feigenbaum, L., Maynard, D., Motta, E., Thirunarayan, K. (eds.) ISWC 2009. LNCS, vol. 5823, pp. 65–80. Springer, Heidelberg (2009)
20. Fowler, M., Beck, K., Brant, J., Opdyke, W., Roberts, D.: Refactoring: Improving the Design of Existing Code. Addison Wesley, Reading (1999)
21. Kanzaki, M.: Exif vocabulary workspace - rdf schema (2003), http://www.w3.org/2003/12/exif/ (last update in 2007)
22. Brickley, D.: Basic Geo (WGS84 lat/long) Vocabulary (2006)
23. Boll, S., Bürger, T., Celma, O., Halaschek-Wiener, C., Mannens, E., Troncy, R.: Multimedia Vocabularies on the Semantic Web (2007)
24. Media Annotations Working Group, http://www.w3.org/2008/WebVideo/Annotations/
25. The Metadata Working Group, http://www.metadataworkinggroup.org/
26. Dasiopoulou, S., Tzouvaras, V., Kompatsiaris, I., Strintzis, M.G.: Enquiring MPEG-7 based multimedia ontologies, pp. 331–370 (2009)

An Approach to Early Recognition of Web User Tasks by the Surfing Behavior

Anne Gutschmidt

University of Rostock, Department of Economic and Organizational Psychology
Ulmenstr. 69, 18057 Rostock, Germany
anne.gutschmidt@uni-rostock.de

Abstract. A study was conducted to investigate Web users' information seeking behavior on online newspapers, distinguishing between the task categories fact finding, information gathering and browsing. Over a period of four weeks, the surfing behavior of 41 users was recorded who additionally kept a diary to document their activities. It was scrutinized whether the surfing behavior shows significant differences depending on the kind of task already at the beginning of an activity, which is a prerequisite for timely reaction to current user needs. According to the results, behavioral aspects, such as the number of pages viewed, scroll and mouse movement behavior etc. produce significant differences already during the first 60 seconds of a task. Nevertheless, classification tests show that these behavioral attributes do not yet lead to a prediction accuracy sufficient for a sound real-time task recognition.

1 Introduction

In the field of Web personalization, especially recommender systems, there are many ways of determining documents and products that may be of use to an individual. While many recommendations are generated based on long-term features such as a user's demographics or product and topic preferences [1], the current context is often neglected. On the other hand, methods which do take into account the current situation, such as association rule mining or sequence analysis, find recommendations based on statistical analysis of users' click paths. A Web page may be recommended to a person because users with a similar click path all ended on this page [2]. However, this is based on correlation and not on causality which means that it is often hard to interpret such personalization mechanisms.

In this paper, a new approach is presented which is similar to market segmentation: Users are assigned to groups that correspond to the kind of task they are currently performing. The task is an important aspect of a user's current context [3]. Moreover, *"[...] a task is the manifestation of an information seeker's problem and is what drives information seeking actions."* [4, p. 36] Consequently, the task reveals what the user needs at the very moment. Based on a present taxonomy of information seeking tasks, comprising the categories fact finding, information gathering and browsing [5], a segmentation is chosen where each

T. Declerck et al. (Eds.): SAMT 2010, LNCS 6725, pp. 64–79, 2011.

segment can be interpreted and where the according needs can be logically and empirically derived.

But before such segmentation can be put into practice, it must be possible to unobtrusively infer a user's current task category. We conducted a study with 41 participants whose surfing behavior was recorded on different online newspapers during four weeks. The participants labeled their surfing activities in a diary. Both the recorded behavior and the label data was used to identify behavioral attributes that show a significant difference depending on the kind of task. However, as task-based personalization requires an early recognition of the task category, we examined only the first 60 seconds of each task recorded. Significance tests pointed out several useful attributes, such as the number of pages viewed and the covered mouse distance. In spite of this selection of useful attributes, classification tests with Naive Bayes led to an insufficient prediction accuracy, pointing out the need for supplementing approaches.

The next section will present related work in this area, followed by a section on the behavioral attributes we created and examined. Section 4 contains a description of the study we conducted. The results of our analysis are presented in section 5 and discussed in section 6. Section 7 summarizes the study's limitations. A final conclusion is given in section 8.

2 Related Work

In marketing, the principle of market segmentation is used to specify groups of customers that have common characteristics, such as demographics, and who accordingly react similarly to product offers, advertising or further marketing instruments [6]. Decades ago, marketing researchers have recognized that the purchase occasion also plays an important role beside stable consumer characteristics. According to [7] half of the variance in consumer preferences can be explained by the situation both alone and in combination with attributes of product and person. On the whole, the situation explains why the same person may react differently to identical product offers on different occasions [8].

The principle of situational segmentation may also be applied in Web personalization by focusing on the question of what kind of task a user is currently performing. The task is an important aspect of the user's current context [3]. According to [9, p. 5], tasks *"are activities that have to be performed to reach a goal"*, thus they are inseparable from the user's current goal and eventually reveal the user's current needs. Having access to this information, personalization can be made even more aimed at a user's presently relevant requirements.

There have been several approaches to categorizing user tasks on the Internet where the categories may be considered as kind of market segments. Marchionini (1989) distinguished between tasks with open-ended and closed questions [10] complemented by a third category called serendipitous browsing [4, 11, 12]. This taxonomy can be seen again in more recent approaches [13, 14, 15, 5].

Kellar et al. introduced a taxonomy of information seeking tasks [5], containing the following three categories:

Fact finding comprises tasks where users are purposefully looking for a specific piece of information, which corresponds to Marchionini's tasks with closed questions.

Information gathering refers to tasks where information on a particular topic is collected, thus open-ended questions trigger this kind of activity.

Browsing relates to serendipitously surfing the Internet without any specific target except for being entertained or generally informed.

The authors added the supplementary categories transactions, such as online purchases, and maintenance, e.g. maintaining web content which is under development [5]. A classification test using decision trees with the four classes fact finding, information gathering, browsing and transactions led to a prediction accuracy of 53.8% [16]. The data, which was collected during a user study, was extremely unbalanced with transactions comprising approximately half of the data examples [5]. The test was based on behavioral features such as task duration, the number of windows opened, the number of pages viewed, browser navigation mechanisms for initiating a new task, the use of browser functions and the use of Google. Individual differences in the navigation style were named as a major source of behavioral variance that eventually produced a low prediction accuracy.

Our approach is to look at the user behavior in more detail and consider attributes such as scroll and mouse speed, clicks on images, clicks on text, etc. such as already presented in [17]. Moreover, we want to test whether the variance in the surfing behavior can be reduced by restricting task recognition to a particular group of Websites. We consider online newspapers as most suitable for an exploratory investigation as they have a common structure and serve the same purpose. Moreover, we want to make a step further towards real-time task recognition and identify attributes that show significant differences depending on the task already at the beginning. The earlier the task can be recognized, the faster a personalization system can react to a user's present requirements.

3 Capturing Web User Behavior

The user behavior can be captured in different ways, depending on the level of detail that is required. The study presented in this paper is based on data collection with a modified browser where every event occurring during surfing is recorded and written into a log file including a time stamp and further information, such as information on the element a mouse click referred to. In contrast to event tracking via proxy server and embedded JavaScript [18], this variant also allows insight into the usage of the browser's graphical user interface [5].

The event logs we collected are a low level representation of the surfing behavior and can be used to generate behavioral attributes at a higher level of abstraction, such as the number of clicks per minute. As we want to find attributes that help distinguishing between the above-presented categories of Web user tasks at the beginning of a task, we created attributes that refer to time slices of ten seconds. In our analysis we only considered the first six of these time slices, thus, we investigated only the first minute of a task. The following sections present the attributes used in our approach, divided by the kind of recorded event they are based on.

3.1 Page Views

Before the field study presented in this paper, a pilot study was conducted to gain an insight into the set of possibly task-discriminating attributes [17]. A lot of these attributes were based on features of the page views, such as the navigation depth, which is expected to reflect how deep a user navigates into a site by considering the number of slashes contained in the URL. We adopted this attribute by assigning the current navigation depth to the respective time slice. Depending on the kind of task, there might be a different point in time when users start to navigate deeper into the Website. During browsing, for example, users will probably stay longer on the start page, indicated by low navigation depth in later time slices.

The analyses from the pilot study have also shown that content-related attributes help in distinguishing between browsing and information gathering. That is why we also considered the number of images contained in all pages viewed until the current point in time, as well as the amount of image area and the number of words. Furthermore, the surfing speed is elicited by considering the number of pages viewed at the end of each time slice. This has already been found useful as a task feature in earlier work [5, 17]. Cumulated values are used for all page view attributes but the navigation depth because general tendencies are to be captured concerning image and text consumption as well as speed.

3.2 Scroll Movement

Scrolling activity has already been considered in the context of deriving a user's interest in a certain document [19]. In our previously conducted pilot study, fact finding was the fastest category concerning scrolling, followed by browsing, with information gathering being the slowest [17]. The field study has to show whether this tendency can be confirmed and whether it already occurs during the first 60 seconds. Four attributes concerning scrolling were generated: the general distance passed by scrolling independently from direction, the distance scrolled up and the distance scrolled down, and the proportion of the current page's length revealed by scrolling (maximum scroll level / page length). All attributes refer only to the current time slice and do not consider previous time slices.

3.3 Mouse Hovering

In former studies, the mouse has been identified as an important device for finding information, as users tend to point on items that seem relevant to them [20]. Moreover, several authors have pointed out a significant relationship between the position of the mouse pointer and a user's eye gaze position [21, 22]. We generated attributes for measuring the number of page items touched by the mouse pointer as well as the amount of time the mouse pointer dwelled on images, text in general, headings in particular, and on empty space for each time slice of ten seconds. The values are not cumulated over time, but refer to the respective time slice, as a development in the course of the task is to be investigated. Is browsing associated with a preference for images throughout the whole task? Do information gathering and fact finding show a preference for text at the beginning of a task which is replaced by a preference for images at a later point in time?

3.4 Mouse Movement

Mouse movement has long been considered as important for estimating a user's interest in a Web document [19, 23]. In the pilot study, fact finding showed the highest mouse speed, whereas information gathering and browsing were similar in speed, although browsing showed longer movement interruptions [17].

For the analysis of time slices, we consider now the amount of pixels the mouse was moved during the time slice. We are especially interested in finding out wether the the suggested task categories show different critical points in time when the mouse movement changes with respect to speed, direction or determination. We separately investigate the amount of pixels moved horizontally and vertically, respectively. In some cases users were found to be using the mouse pointer to support reading by moving it over text lines from left to right [24, 25]. We hypothesized that information gathering would involve more reading than the other task categories.

Similar to performance measures of pointing tasks such as presented in [26], we considered a simple measure, 'detour' which is the proportion of the mouse distance covered and the actual length of the straight line between the starting point of the mouse and the final point in the respective time slice. We expected to find that the task categories are different in the way users more or less purposefully move the mouse pointer.

4 Diary Study

4.1 Study Goal

The goal of the study was to investigate the relationship between the surfing behavior and the user task. The event logs offer a detailed image of the user behavior and we wanted to find out which behavioral attributes really contribute to the recognition of the task category. Based on a selection of task-discriminating attributes, we want to find out whether a sufficient prediction accuracy can be attained using common machine learning methods such as Naive Bayes classification. However, the particular goal of this approach is to consider only the first 60 seconds of a task and find out whether this time period is already enough to bring about useful behavioral attributes that may lead to sound task recognition.

4.2 Sample

We recruited 41 students. The majority studied computer science and business administration and took part in a computer science class. On average, the participants were 21.5 years old, 10 of them were female and 31 male.

4.3 Tracking News Reading

As the investigations focused on the use of online newspapers, the event tracking software was prepared in such a way that recording was switched on as soon as the participants entered a predefined site, and switched off as soon as the site was left. This was done to avoid unnecessarily intruding upon the participants' privacy. The tracking software was enhanced with a surfing diary, presented as a separate dialog window, where the test persons documented their activities by choosing a task label. They could choose between fact finding, information gathering, browsing, and transactions, other and no idea. With 'other' we wanted to give the participants an opportunity to label tasks of which they were sure they would not belong to any of the other tasks. 'No idea' was recommended to be chosen when they were just unsure which category to choose. Additionally, all tasks had to be described in keywords concerning target and occasion, however, only the category label is used for the analysis presented in this paper. Several ways of facilitating the documentation of the tasks were integrated in the diary: By clicking on a particular page view, details such as title and URL are displayed to the user. Moreover, by clicking on a category label, the associated explanation is given.

4.4 Setting

The participants were invited in groups to an inaugural meeting where the general course of the study was explained. They were made familiar with the tracking software including the diary module. Moreover, the task taxonomy by Kellar et al. was introduced and practiced in short exercises. After this meeting the participants installed the tracking software on their computers which recorded the surfing behavior on a predefined list of German online newspapers, such as *www.spiegel.de* or *www.focus.de*, during a period of four weeks. During this period, the participants were repeatedly motivated to read online newspapers by placing animating links or tasks on the browser's start page, such as looking up the weather forecast of next weekend. After half of the study's duration, the participants had to complete a test where their knowledge on the task categorization was refreshed.

5 Results

The data that was analyzed comprised 362 task data sets of which 29 belonged to fact finding (FF), 140 to information gathering (IG), and 193 to browsing. The task categories 'transactions' and 'other' were not considered due to their very low number. On the whole, only two tasks were labeled as transactions, which is surprising facing the increasing importance of user participation such as with blogs and commenting [27]. The reason for this might be that established mainstream newspapers like those we have considered in the study are not a typical place of social exchange for users of such young age.

Time slices were determined by dividing the tasks into periods of ten seconds. Most time slices belonged to the first page view. If a remainder shorter than ten seconds occurred at the end of a page view, it was excluded. Time between page views, for example during waiting for the next page to be loaded, was not considered, because the data should refer to the users' interaction with completely loaded pages. Of each task, the first six time slices of ten seconds were analyzed; i.e. for each of the six time slices 362 data sets were evaluated that were described by the behavioral attributes presented in section 3. Only tasks longer than 60 seconds were included, as equal sample size in all time slices was desired and many tasks did not reach a duration longer 60 seconds.

The selected time slices were examined with respect to significant differences of behavioral attributes produced by the respective task category, as presented in section 5.1. Moreover, classification tests were performed with a Naive Bayes classifier and are presented in section 5.2.

5.1 Examination of Attributes

To find significant differences between the task categories, the Kruskal-Wallis test was performed on the behavioral attributes in each time slice. The Kruskal-Wallis test is a non-parametric test that does not require normal distribution and homoscedasticity which are not given for most of the attributes considered here [28]. Table 1 contains average values and standard deviations for every attribute and time slice where a significant difference has been found by the Kruskal-Wallis test. Non-significant attributes were omitted. The two right-hand side columns belong to the Kruskal-Wallis test where $p <= 0.05$ was chosen as significance threshold. Fig. 1 shows the development of the average value of exemplary attributes over time.

While the Kruskal-Wallis test shows a global difference for the respective attribute, differences between concrete pairs of task categories are pointed out with the help of Mann-Whitney tests. Similar to Kruskal-Wallis, Mann-Whitney treats the data as ordinal and may thus be less accurate than a t-test. However, in case of missing normal distribution and homogeneity of variance it is to be preferred over parametric tests [29]. In Table 2, significant results are marked with a star * where the significance threshold α was set to $p <= 0.05$. An adaptation of this threshold, which is usually applied to prevent Type I errors when testing concrete hypotheses, was not performed; e.g. with Bonferroni correction the usual threshold value is divided by the number of tests which can diminish the test power immensely [30, 31]. In exploratory examinations like this one, it is admissible to prefer a greater statistical power, as hypotheses are still to be formulated.

Table 1. Results of the Kruskal-Wallis test for all newspapers, $df = 2$

		FF		IG		B		K.-W. test	
Slice	Attribute	μ	σ	μ	σ	μ	σ	χ^2	p
I: 1-10	WordNumber $\times 10^3$	3.96	1.98	4.05	2.13	7.15	7.75	9.074	0.011
II: 11-20	PvNumber	1.41	0.57	1.46	0.84	1.30	0.74	6.357	0.042
	MaxScrollY %	24.30	12.80	36.16	28.86	29.74	25.05	6.119	0.047
	MouseDistance px $\times 10^3$	0.64	0.63	0.48	0.69	0.40	0.66	7.219	0.027
	MouseDistanceX px $\times 10^3$	0.44	0.51	0.34	0.54	0.29	0.53	6.612	0.037
	MouseDistanceY px $\times 10^3$	0.30	0.32	0.23	0.33	0.19	0.32	7.819	0.020
III: 21-30	PvNumber	2.00	1.07	1.70	1.14	1.52	1.04	8.958	0.011
	MouseDistanceX px $\times 10^3$	0.37	0.42	0.30	0.49	0.27	0.53	6.015	0.049
IV: 31-40	PvNumber px $\times 10^3$	2.45	1.40	1.93	1.30	1.75	1.21	9.110	0.011
	WordNumber $\times 10^3$	7.15	4.60	5.90	3.69	10.09	12.16	6.379	0.041
	ScrollUp px $\times 10^3$	0.10	0.18	0.35	1.20	0.08	0.31	7.989	0.018
	MouseDistanceY px $\times 10^3$	0.27	0.35	0.26	0.45	0.16	0.35	6.301	0.043
V: 41-50	PvNumber	2.76	1.38	2.22	2.17	1.93	1.36	12.193	0.002
	WordNumber $\times 10^3$	7.87	4.77	6.42	3.98	10.87	12.65	6.402	0.041
VI: 51-60	PvNumber	3.21	1.40	2.47	2.44	2.11	1.54	18.197	0.000
	WordNumber $\times 10^3$	9.32	5.67	6.99	4.32	11.74	14.21	7.031	0.030
	ScrollUp px $\times 10^3$	0.32	0.69	0.27	0.80	0.18	0.56	6.534	0.038
	MouseHoverNumber	6.34	7.50	4.00	5.97	3.55	6.46	6.036	0.049
	MouseHoverHead s	0.15	0.47	0.18	0.91	0.11	0.62	13.063	0.001
	MouseDistance px $\times 10^3$	1.01	1.33	0.59	0.92	0.36	0.63	10.432	0.005
	MouseDistanceX px $\times 10^3$	0.56	0.72	0.40	0.64	0.25	0.45	9.489	0.009
	MouseDistanceY px $\times 10^3$	0.63	0.99	0.31	0.56	0.19	0.37	11.001	0.004
	Detour	8.44	24.28	4.63	20.07	2.74	6.87	8.716	0.013

The results in Table 2 show that the first slice has the lowest number of task-discriminating attributes: only one task-dependent attribute, the number of words that was presented to the user until the tenth second. Slice six contains the greatest number of task-discriminating attributes. As can be seen, most of the attributes appear in several time slices, such as the number of page views and the number of words presented.

5.2 Automatic Task Prediction

The task-discriminating attributes found with the significance tests were used as a pre-selection of classification features. Furthermore a correlation filter removing highly correlating attributes (Pearson correlation >= 0.6) was applied. Subsequently, a classification with Naive Bayes including forward feature selection and ten-fold cross-validation was performed on the attributes, for each time slice in separate. The upper part of Table 3 shows the detailed results using the whole data set with 362 examples. As the data is extremely unbalanced with a very high number of examples for browsing and a very low number of fact finding examples, the analysis was repeated using balanced samples: 29 examples

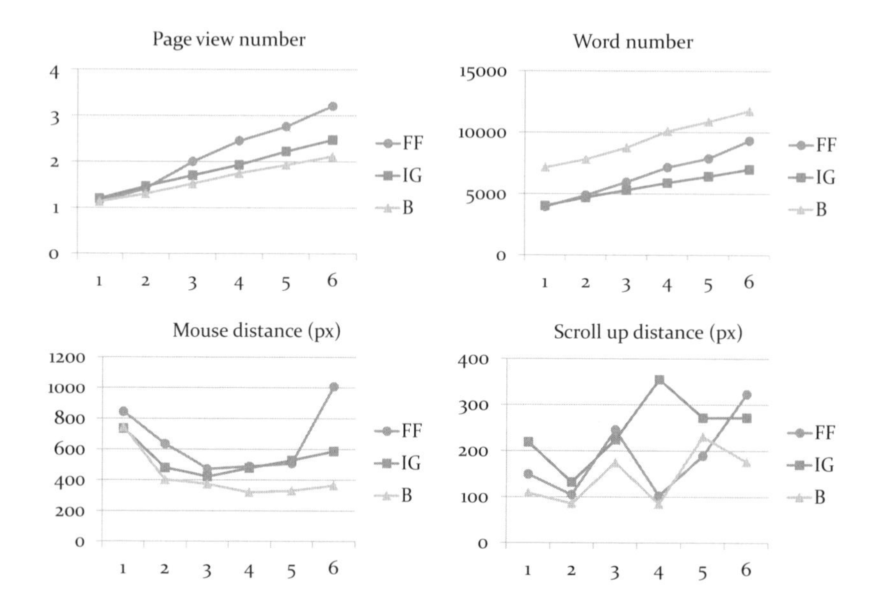

Fig. 1. Development of average values over six time slices for selected attributes, the x-axis refers to the slice index, the y axis refers to the average value

were randomly chosen from information gathering and browsing, respectively, and put together with the 29 original fact finding examples. The results using the balanced data are contained in the lower part of Table 3. Table 3 presents the overall prediction accuracy as well as precision and recall values for each class or task category, respectively. Precision represents the proportion of examples where a particular class was predicted and to which they indeed belonged. Class recall is the number of examples belonging to a class that have been identified as such in proportion to the class size. The development of the precision values based on the balanced data is also depicted in Fig. 2.

6 Discussion

Looking at the distribution of task examples among the three categories, it becomes obvious that online newspapers are a typical target for browsing. More than half of the tasks were labeled as browsing by the participants whereas fact finding comprises less than 10% of the tasks.

Table 1 illustrates the fluctuating number of task-discriminating behavioral attributes over time. What sticks out the most is that during the first ten seconds there is the lowest number of significantly differing attributes. The time slice with the highest number of useful attributes is the sixth and last, with nine task-discriminating attributes. Thus it seems, as one might already have expected, that the more time passes the clearer the differences between the three task categories become.

Table 2. Results of the Mann-Whitney test

		FF – IG		FF – B		IG – B	
Slice	Attribute	Z	p	Z	p	Z	p
I: 1-10	WordNumber	-0.390	0.697	-1.949	0.051	-2.693	0.007*
II: 11-20	PvNumber	-0.447	0.655	-1.903	0.057	-2.172	0.030*
	MaxScrollY	-2.031	0.042*	-1.180	0.238	-1.905	0.057
	MouseDistance	-1.819	0.069	-2.538	0.011*	-1.365	0.172
	MouseDistanceX	-1.845	0.065	-2.541	0.011*	-1.067	0.286
	MouseDistanceY	-1.618	0.106	-2.540	0.011*	-1.691	0.091
III: 21-30	PvNumber	-1.935	0.053	-2.996	0.003*	-1.362	0.173
	MouseDistanceX	-1.636	0.102	-2.246	0.025*	-1.348	0.178
IV: 31-40	PvNumber	-2.243	0.025*	-3.104	0.002*	-0.931	0.352
	WordNumber	-1.293	0.196	-0.175	0.861	-2.463	0.014*
	ScrollUp	-1.167	0.243	-2.580	0.010*	-2.012	0.044*
	MouseDistanceY	-1.696	0.090	-2.618	0.009*	-0.905	0.365
V: 41-50	PvNumber	-2.710	0.007*	-3.646	0.000*	-0.788	0.431
	WordNumber	-1.664	0.096	-0.091	0.927	-2.359	0.018*
VI: 51-60	PvNumber	-3.334	0.001*	-4.437	0.000*	-0.979	0.328
	WordNumber	-2.152	0.031*	-0.512	0.609	-2.271	0.023*
	ScrollUp	-2.126	0.033*	-2.569	0.010*	-0.295	0.768
	MouseHoverNumber	-1.409	0.159	-2.140	0.032*	-1.591	0.112
	MouseHoverHead	-2.224	0.026*	-3.643	0.000*	-1.739	0.082
	MouseDistance	-1.762	0.078	-2.951	0.003*	-1.990	0.047*
	MouseDistanceX	-1.558	0.119	-2.860	0.004*	-1.915	0.055
	MouseDistanceY	-1.843	0.065	-2.982	0.003*	-2.076	0.038*
	Detour	-1.581	0.114	-2.715	0.007*	-1.816	0.069

Table 3. Classification results for unbalanced and balanced data, by time slice

			Precision			Recall		
	Slice	Accuracy	FF	IG	B	FF	IG	B
unbalanced	1	51.10%	0.00%	44.21%	76.62%	0.00%	90.00%	30.57%
	2	53.34%	0.00%	50.00%	53.68%	0.00%	12.86%	90.67%
	3	54.16%	0.00%	52.00%	54.46%	0.00%	9.29%	94.82%
	4	53.89%	0.00%	51.16%	54.57%	0.00%	15.71%	89.64%
	5	55.56%	0.00%	52.50%	56.07%	0.00%	15.00%	93.26%
	6	52.21%	0.00%	38.30%	54.98%	0.00%	12.86%	88.60%
balanced	1	43.75%	37.50%	42.86%	80.00%	72.41%	31.03%	27.59%
	2	43.75%	42.42%	61.54%	39.02%	48.28%	27.59%	55.17%
	3	43.89%	44.00%	42.86%	43.75%	37.93%	20.69%	72.41%
	4	47.22%	50.00%	41.38%	69.23%	27.59%	82.76%	31.03%
	5	50.69%	59.09%	43.64%	70.00%	44.83%	82.76%	24.14%
	6	49.44%	59.26%	44.44%	50.00%	55.17%	82.76%	10.34%

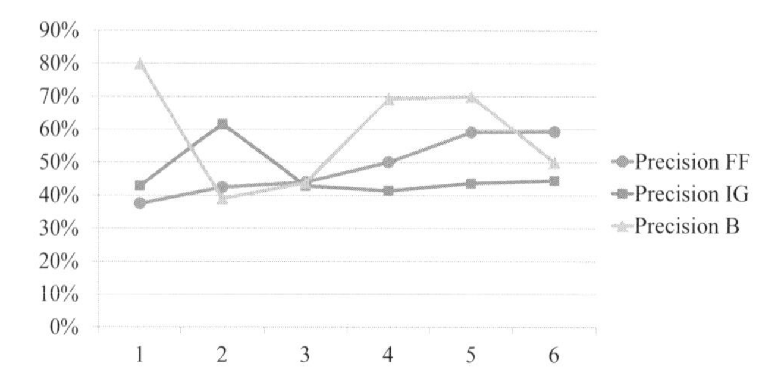

Fig. 2. Development of the class precision over time, based on balanced samples

The standard deviations contained in Table 1 point out the high variance in the users' surfing behavior in spite of the fact that a group of kindred Websites has been chosen. Consequently, individual habits and styles still seem to have a strong influence.

Regarding Table 1 and Table 2 in combination, several assumptions can be derived for each attribute in each time slice.

Page view attributes: With respect to the navigation depth, all task categories did not seem to cause a different behavior during the first 60 seconds. The users showed the same gradual increase in navigation depth with similar attribute values in all tasks. Page view features concerning the presentation or consumption, respectively, of images did neither lead to behavioral differences. Consequently, a preference for images was not found. However, significant differences concerning the amount of words presented occurred over time. As the values for 'wordNumber' in Table 2 indicate, information gathering and browsing were significantly different in time slices one, four, five and six. The chart contained in Fig. 1 illustrates that browsing seems to be generally associated with Web pages containing more text. A further significant difference for this attribute was found in time slice six between fact finding and information gathering due to an increase of the attribute's values in fact finding. All in all, there is no task difference with respect to image consumption, but browsing seems to be associated with viewing pages with more text. Probably, the users spent more time on newspapers' front pages because these are often longest and contain the greatest amount of text. Measuring the number of pages viewed (pvNumber) at the six different points in time makes clear that fact finding tasks are generally performed faster (see chart in Fig. 1). The users clicked through more pages during the first 60 seconds, indicated by significant differences between fact finding and browsing from slice three until slice six, and between fact finding and information gathering from slice four to slice six. Information gathering tasks and browsing tasks are generally equal in surfing speed, only in slice two a significant difference was found where information gathering appeared quicker. The speed finally explains why wordNumber increases for fact finding in slice six.

Scroll Movement: There were no significant differences found between the categories in regard to general scrolling and scrolling down in particular. Concerning scrolling upwards, however, there was a difference between information gathering and browsing in slice four and between fact finding and browsing in slices four and six, where the difference in slice four seems very narrow in Fig. 1. This might be caused by the fact that the Mann-Whitney test is less accurate than a t-test as it treats the data as ordinal data. However, as explained above, due to violated preconditions of the respective parametric tests, it is to be considered as more suitable. When a user scrolls upwards he or she probably intends to go back to the navigation menu situated on top of each page or to go back to to upper parts of the page to find something. In slice four, information gathering seems to cause such behavior more than the other categories, as illustrated in the chart presented in Fig. 1. In slice six, scrolling up is more prominent in fact finding. With respect to the maximum scroll level, the results show a significant difference only for fact finding and information gathering in slice two. At this point in time, information gatherers have scrolled down further already. All in all, the task categories show a similar scroll behavior with few significant differences.

Mouse Hovering: No differences were found when considering the task categories with respect to pointing the mouse on images, general text or empty space. Concerning the number of page items touched by the mouse pointer, there was also no significant difference between the task categories during the first 50 seconds. However, in slice six, it seems that fact finders become faster with the mouse as more page items are touched, although a significant difference could only be found in comparison with browsing. This high number of touched page items indicates that the users did not point long on all the items, but probably only grazed most of them while moving the mouse quickly over the page. In slice six, the task categories also show differences in pointing on headings: during information gathering, the users seem to hover the mouse longer over headings, whereas during browsing, the mouse dwells the least time on headings. Fact finding is situated in between the other two classes at this point in time. However, the difference between information gathering and browsing is not significant according to the Mann-Whitney test.

Mouse Movement: Considering the distance passed with the mouse pointer, fact finding appears to be associated with fast mouse movement, whereas browsing showed the slowest mouse movement, as also depicted in the chart contained in Fig. 1. Significant differences were found in general mouse movement between fact finding and browsing in slices two and six and between information gathering and browsing in slice six. Further differences were found in vertical and horizontal mouse movement, where fact finding and browsing were clearly separated, as significant differences were found in three time slices for both attributes. Again fact finding appeared significantly faster. However, the results for vertical and horizontal mouse movement do not give significant hint on reading activity. With respect to the 'detour' attribute, fact finding showed a significant tendency towards passing more additional way with the mouse pointer in slice six.

However, the Mann-Whitney test showed only a difference between fact finding and browsing. Nevertheless, the results confirm that fact finding seems to be associated with wasteful and fast mouse movement.

To sum up, the following behavior profiles can be derived for the three examined task categories during the first 60 seconds: **Fact finding** is performed quickly, expressed by the significantly greater mouse distance passed, a high number of page items grazed, especially in the sixth time slice, and the high number of pages viewed. **Information gatherers** seem to be purposeful and efficient in searching, as they tend to scroll pages far down and up again early, probably to quickly find the information they need. Their significant interest in headings in slice six might also give hint on a concentrated proceeding. Concerning mouse movement, the users showed a medium speed. **Browsing** seems to be associated with a preference for Web pages with a lot of text. The users showed less mouse movement compared to the other two categories. This might give hint that during browsing, users are more concentrated on the content and probably on reading text, whereas the behavior of fact finding and information gathering may be mostly influenced by the intention of searching something specific.

In the original data set, fact finding appears with a probability of approximately 8% which makes the Naive Bayes classifier rarely choose this label. On the other hand, there are so many browsing examples that the probability of correctly labeling an example as browsing is very high. Due to this imbalance, the time slices' classification results are presented in Table 3 both with unbalanced and balanced data where the latter should give a better insight into the prediction performance. Using balanced samples, the prediction accuracy remains stable during the first 30 seconds, at the fourth slice it increases until the end of the fifth slice. However, the maximum accuracy of 50% is to be considered as insufficient. Although a group of similar Websites was considered and the study sample was relatively homogeneous in age and Internet background, there was a considerable variance in the user behavior which is an important finding in this study.

Nevertheless, the detailed precision and recall values over time show that the approach presented is still promising. Fig. 2 shows the development of precision for the three categories over time. During the first ten seconds a prediction of browsing is correct with a probability of 80%. Information gathering reaches a maximum precision after 20 seconds with 61%, and fact finding after 60 seconds with 59%.

This could be a basis for extending this approach by considering the development of the prediction performances of the three categories over time. However, a major condition is that the noise caused by individual surfing habits must be either elicited and considered, e.g. by recognizing the surfing type, or further parameters have to be found, e.g. based on page content, time of day etc., that surmount this noise.

7 Limitations

The study presented in this paper served as an exploratory investigation of the relationship between surfing behavior and task category for which the given sample and its size can be considered as sufficient. However, the results, especially the conclusions concerning the usefulness of various behavioral attributes have to be tested with an extended and more heterogeneous sample in future studies. Moreover, it is desirable to perform further analyses on other types of Websites where tasks referring to transactions can be considered in addition. However, the above-mentioned limitations are also a strength of the study, because this way it was possible to show that there are more factors of disturbance than the kind of Website or age or Internet experience which significantly cause variance in the user behavior.

One particular limitation of the study is that the results are significantly based on the participants' willingness and ability to label their surfing activities during the whole study. We tried to maintain the test persons' motivation by showing interesting questions and links on the browser's start page. A test in the middle of the study was to refresh the participants' knowledge on the task taxonomy. Furthermore, a reminder functionality was integrated in the diary software to make the users document the tasks timely. As another limitation one may criticize that the use of the Firefox browser was required. However, 34 of the 41 participants stated to be regular Firefox users.

Finally, it must be noted that we present a work still in progress. Examinations concerning the most suitable size of time slice as well as the determination of the minimum duration to recognize the task category are still to follow. Furthermore, there are certainly more machine learning techniques that must be tested on the data. We intend to perform further classification tests in the near future based on methods which can handle the challenge of unbalanced data.

8 Conclusion

An exploratory study on the relationship between user behavior and user task was conducted. Based on event logs, several behavioral attributes were generated to identify the kind of information seeking task a user is performing on online newspapers: fact finding, information gathering or browsing. The analyses concentrated on the first 60 seconds of surfing, as an early recognition of the task category is a precondition for timely personalization. Significance tests were used to select those attributes that show a significant difference depending on the task within a time slice of ten seconds, such as the number of pages viewed or the number of words presented on previously viewed pages. Predicting the task category using a Naive Bayes classifier does not yet lead to a sufficient accuracy as the behavioral variance caused by individual surfing styles produces considerable noise. However, the three categories show promising precision values in the classification over time; e.g., after ten seconds, browsing is predicted with 80% precision. Thus, the approach presented in this paper may be seen as a good starting point for further research.

In particular, the approach goes beyond preceding work, as the behavior is considered in more detail and not only the task as a whole has been investigated, but its first 60 seconds, making a step further towards real-time task recognition. Several behavioral attributes were presented that may also be applied in other fields such as usability analysis. Moreover, it was shown that there exist significant differences in the user behavior which are independent on the kind of Website as well as on age or Internet experience. Future studies will have to explore the aspect of individual surfing styles more deeply.

Finally, knowing what kind of task a user is currently performing is the key to more context and semantics-driven personalization. Each task category may be interpreted and understood, similar to a market segment. Thus, personalization strategies can be deliberately chosen and would no longer be based on correlations only. This makes the recognition of the task category such a desirable target.

References

[1] Mobasher, B.: Data mining for web personalization. The Adaptive Web, 90–135 (2007)

[2] Das, A., Datar, M., Garg, A.: Google news personalization: Scalable online collaborative filtering. In: Proceedings of the 16th International Conference on World Wide Web, pp. 271–280. ACM Press, New York (2007)

[3] Brusilovsky, P., Millán, E.: User models for adaptive hypermedia and adaptive educational systems. The Adaptive Web, 3–53 (2007)

[4] Marchionini, G.: Information seeking in electronic environments. Cambridge University Press, Cambridge (1997)

[5] Kellar, M., Watters, C., Shepherd, M.: A field study characterizing web-based information-seeking tasks. Journal of the American Society for Information Science and Technology 58(7), 999–1018 (2007)

[6] Beane, T., Ennis, D.: Market segmentation: a review. European Journal of Marketing 21(5), 20–42 (1987)

[7] Belk, R.: An exploratory assessment of situational effects in buyer behavior. Journal of Marketing Research 11(2), 156–163 (1974)

[8] Hall, J., Lockshin, L.: Using means-end chains for analysing occasions-not buyers. Australasian Marketing Journal 8(1), 45–54 (2000)

[9] Paterno, F.: Model-based design and evaluation of interactive applications (1999)

[10] Marchionini, G.: Information-seeking strategies of novices using a full-text electronic encyclopedia (1989)

[11] Catledge, L.D., Pitkow, J.E.: Characterizing browsing strategies in the world-wide web. Computer Networks and ISDN systems 27(6), 1065–1073 (1995)

[12] Cove, J.F., Walsh, B.C.: Online text retrieval via browsing. Information Processing & Management 24(1), 31–37 (1988)

[13] Rozanski, H.D., Bollman, G., Lipman, M.: Seize the occasion - usage-based segmentation for internet marketers. Technical report, Booz Allen & Hamilton (2001)

[14] Morrison, J.B., Pirolli, P., Card, S.K.: A taxonomic analysis of what world wide web activities significantly impact people's decisions and actions. In: CHI 2001 Extended Abstracts on Human Factors in Computing Systems, pp. 163–164. ACM, New York (2001)

[15] Sellen, A.J., Murphy, R., Shaw, K.L.: How knowledge workers use the web. In: CHI 2002: Proceedings of the SIGCHI Conference on Human Factors in Computing Systems, pp. 227–234. ACM, New York (2002)

[16] Kellar, M., Watters, C.: Using web browser interactions to predict task. In: WWW 2006: Proceedings of the 15th International Conference on World Wide Web, pp. 843–844. ACM Press, New York (2006)

[17] Gutschmidt, A.: The prediction of web user tasks by analyzing client logs. IADIS International Journal on WWW/Internet 6 (2009)

[18] Atterer, R., Wnuk, M., Schmidt, A.: Knowing the user's every move: user activity tracking for website usability evaluation and implicit interaction. In: Proceedings of the 15th International Conference on World Wide Web, WWW 2006, pp. 203–212. ACM, New York (2006)

[19] Claypool, M., Le, P., Wased, M., Brown, D.: Implicit interest indicators. In: Proceedings of the 6th International Conference on Intelligent User Interfaces, IUI 2001, pp. 33–40. ACM, New York (2001)

[20] Cox, A., Silva, M.: The role of mouse movements in interactive search. In: Proceedings of the 28th Annual CogSci. Conference, pp. 26–29 (2006)

[21] Chen, M.C., Anderson, J.R., Sohn, M.H.: What can a mouse cursor tell us more? correlation of eye/mouse movements on web browsing. In: Conference on Human Factors in Computing Systems, CHI 2001 Extended Abstracts on Human Factors in Computing Systems, pp. 281–282 (2001)

[22] Mueller, F., Lockert, A.: Cheese: Tracking mouse movement activity on websites, a tool for user modeling. In: Conference on Human Factors in Computing Systems, CHI 2001 Extended Abstracts on Human Factors in Computing Systems, pp. 289–290 (2001)

[23] Goecks, J., Shavlik, J.: Learning users' interests by unobtrusively observing their normal behavior. In: Proceedings of the 5th International Conference on Intelligent User Interfaces, pp. 129–132. ACM, New York (2000)

[24] Rodden, K., Fu, X., Aula, A., Spiro, I.: Eye-mouse coordination patterns on web search results pages. In: CHI 2008 Extended Abstracts on Human Factors in Computing Systems, pp. 2997–3002. ACM, New York (2008)

[25] Arroyo, E., Selker, T., Wei, W.: Usability tool for analysis of web designs using mouse tracks. In: CHI 2006 Extended Abstracts on Human Factors in Computing Systems, pp. 484–489. ACM, New York (2006)

[26] MacKenzie, I., Kauppinen, T., Silfverberg, M.: Accuracy measures for evaluating computer pointing devices. In: Proceedings of the SIGCHI Conference on Human Factors in Computing Systems, pp. 9–16. ACM, New York (2001)

[27] Rasmussen, S.: News as a service: Adoption of web 2.0 by online newspapers. In: Management of the Interconnected World: ItAIS: the Italian Association for Information Systems, pp. 11–19 (2010)

[28] Dupont, W.: Statistical modeling for biomedical researchers: a simple introduction to the analysis of complex data. Cambridge University Press, Cambridge (2002)

[29] Black, K.: Business statistics: Contemporary decision making. Wiley, Chichester (2009)

[30] Cohen, J.: Statistical power analysis. Current Directions in Psychological Science 1(3), 98–101 (1992)

[31] Nakagawa, S.: A farewell to bonferroni: the problems of low statistical power and publication bias. Behavioral Ecology 15(6), 1044–1045 (2004)

Mapping Audiovisual Metadata Formats Using Formal Semantics

Martin Höffernig, Werner Bailer, Günter Nagler, and Helmut Mülner

JOANNEUM RESEARCH Forschungsgesellschaft mbH
DIGITAL – Institute for Information and Communication Technologies
Steyrergasse 17, 8010 Graz, Austria
firstname.lastname@joanneum.at

Abstract. Audiovisual archives hold enormous amounts of valuable content. However, many of these archives are difficult to access, as their holdings are documented using a range of different metadata formats. Being able to exchange metadata is the key to ensuring access to these collections and to establish interoperability among audiovisual collections, between audiovisual collections and other cultural heritage institutions as well as portals such as Europeana. In this work we attempt to model mappings between metadata formats based on a high-level intermediate concept representation in order to avoid hand-crafted one-to-one mappings between metadata formats. In addition, we define mapping templates on data type level, from which the code for mapping instructions between a pair of formats can be derived. The high-level intermediate concept representation is based on the existing meon ontology and the resulting mapping instructions are expressed using XSLT. As a proof of concept, mappings between two metadata formats are formalized and integrated in a web based prototype application.

1 Introduction

There are millions of hours of audiovisual content in collections of dedicated audiovisual archives, other archives, libraries and museums. Preserving digital audiovisual content means keeping it accessible and usable for a long time, including the metadata describing it. This requires migrating metadata to models currently in use and enabling metadata mapping to formats used in the systems which need to retrieve, deliver, display and process archived content. Being able to exchange metadata is the key to ensuring access to audiovisual collections, establishing interoperability among audiovisual collections, and between audiovisual collections and other cultural heritage institutions. Metadata exchange is often hindered by the diversity of metadata formats and standards that exist in the media production process and in different communities.

The work described in this paper builds on an existing multimedia ontology used for mapping metadata concepts in audiovisual media production workflows. The contributions of this paper are the extension of this ontology to include format-specific concepts and the support for data type conversions in the mapping system.

T. Declerck et al. (Eds.): SAMT 2010, LNCS 6725, pp. 80–94, 2011.

The rest of this paper is organized as follows. The remainder of this section discusses the motivation for developing a novel mapping approach and analyzes the problem and the resulting requirements. Section 2 discusses related work. In Section 3 we propose our approach for mapping and demonstrate a prototype implementation in Section 4. Section 5 concludes the paper.

1.1 Motivation

There is a number of use cases in the audiovisual archive domain that require metadata mapping. On one side these use cases include in-house scenarios such as the conversion of legacy technical metadata as part of preservation processes, access to legacy content descriptions in order to map between material and editorial entities, or the extraction metadata embedded in file headers and convert it to the data structures needed for import into a preservation system. On the other side there are use cases across institutions, such as content exchange with other cultural heritage institutions, content provision to access portals (e.g. Europeana[1]) and outsourcing of annotation and access services, with potentially different data models between the archive's and service provider's infrastructure.

Tools for metadata mapping are needed to overcome the existing interoperability issues on both syntactic and semantic level. However, with n formats existing in a given environment, we need in the worst case $O(n^2)$ mappings if we go for a simple approach considering only pair-wise mappings. Chaining mappings is also not a useful approach, as transitivity of relations cannot be ensured due to the incompleteness of mappings. We thus propose an approach that uses a high-level intermediate representation, together with mapping templates on data type level, from which the code for a mapping problem between a pair of standards can be derived. This would ideally allow us to solve the problem with $O(2n)$ definitions.

1.2 Mapping Problem and Resulting Requirements

The metadata formats encountered across the audiovisual archive domain differ in various aspects.

Coverage. MPEG-7 [10] for example aims to be domain independent while TV Anytime [7] focuses on broadcast metadata for consumers.

Comprehensiveness. For example, MPEG-7 aims to provide comprehensive descriptions of multimedia content ranging from low-level features that can be extracted automatically to fine-grained semantic description of a scene, while the Dublin Core Element Set[2] provides a simple list of general annotation elements.

Complexity. Metadata formats also differ in the complexity of their description syntax. For example, the Dublin Core `dc:creator` element is a simple name or an URI identifying an agent whereas the creator's name in MPEG-7 is

[1] `http://www.europeana.eu`

[2] `http://dublincore.org/documents/dces/`

divided into a complex nested structure of titles, family names and given names along with the definition of her role in the creation process.

The problem of metadata interoperability exists on two levels:

Syntactic interoperability. Even if metadata is represented and processed in the same syntactic format, today typically some XML-based format, syntactic interoperability issues concerning e.g. data type representations and different structural granularity still exist. Note that this does not imply that all metadata formats need to be defined as XML documents, only that they can be rendered as such (with services or wrappers).

Semantic interoperability. Metadata elements in different formats are often not fully congruent w.r.t. to their semantics, but are defined to be narrower or wider than the related elements in other formats. In addition, some formats leave room for interpretation, so that different organization will use the elements in a different way. Also, controlled vocabularies can be used for the values of several elements, thus mapping requires alignment of (possibly in-house) vocabularies.

We define the following requirements for our mapping tool. Given an input metadata document in some supported, XML represented format the mapping tool must provide a best practice mapping of the metadata document to another supported, XML represented target format. The target document must be valid w.r.t. to the standards and must convey the semantics of the input document as far as possible. For each supported format, it is assumed that the mapping tool has access to the following definitions:

- A formal description of the set of system-wide supported metadata elements and their relations. This is independent of a format.
- A formal description of the set of metadata elements in the format and their relations among them (if applicable) and their relations to the set of elements supported by the system.
- A set of mapping rules to be applied.

The mapping tools should be deployed as services in order to allow integration into a wide range of components (which may be located remotely).

There are a number of foreseen difficulties and limitations:

- For some pairs of formats, mapping will be incomplete and lossy in one direction, and will have only loosely defined semantics (or need appropriate additional rules justified by the use case) in the opposite direction. This also means, that bidirectional mapping for such a pair of standards is only possible in a very limited way.
- There are ambiguities in mappings that need default or use case specific rules to be resolved.
- If formats use specific data types, their mappings to (structures of) standard data types needs to be defined. This will reduce the saving in the number of mapping definitions needed.

2 Related Work

There are different approaches described in literature for the problem of mapping between different metadata models, mostly focusing on XML-based representations. Xing et al. [12] present a system for automating the transformation of XML documents using a tree matching approach. However, this method has an important restriction: the leaf text in the different documents has to be exactly identical. This is hardly the case when combining different metadata standards. Likewise, Yang et al. [13] propose to integrate XML Schemas. They use a more semantic approach, using the ORA-SS data model to represent the information available in the XML Schemas and to provide mappings between the different documents. The ORA-SS data model allows to define objects and attributes to represent hierarchical data, however, more advanced mappings involving semantic relationships cannot be represented.

Cruz et al. [2] introduced an ontology-based framework for XML semantic integration. For each XML source integrated, a local RDFS ontology is created and merged in a global ontology. During this mapping, a table is created that is further used to translate queries over the RDF data of the global ontology to queries over the XML original sources. The authors assume that every concept in the local ontologies is mapped to a concept in the global ontology. This assumption can be hard to maintain when the number and the degree of complexity of the incorporated ontologies increases. Poppe et al. [11] advocate a similar approach to deal with interoperability problems in content management systems. An OWL upper ontology is created and the different XML-based metadata formats are represented as OWL ontologies and mapped to the upper ontology using OWL constructs and rules. However, the upper ontology is dedicated to content management system and, as such, is not as general as the approach proposed in this paper.

In [9] the authors propose a format for describing mappings between data models in the domain of database schema conversion. They define a very simple format expressed as a XML DTD which can contain mapping definitions. They suggest a graphical tool to make the process more intuitive. Of course mapping database schemas is a subset of the problem of mapping XML schemas. The approach covers a set of common heterogeneity issues encountered during mapping, to which extensions and exceptions can be defined when needed. The common cases covered by the mapping language are introduction of intermediate nodes, contraction of compounds, transformation from parallel to nested structures (including a specific case with an intermediate node), using the same instance in multiple mappings and conditional mappings.

The problem of mapping between different multimedia metadata formats is also encountered in different standardization initiatives. JPSearch[3] is a project issued by the JPEG standardization committee to develop technologies that enable search and retrieval capabilities among image archives, consisting of five parts. Part 2 introduces an XML based core metadata schema and transformation rules

[3] http://www.jpsearch.org

for mapping descriptive information (e.g., core metadata to MPEG-7 or core metadata to Dublin Core) between peers [4], and part 3 adapts a profile of the MPEG Query Format [5] for ensuring standardized querying. The focus of this approach is clearly querying, thus the approach is concerned with mapping source format to a common representation, but not to map between an arbitrary pair of formats.

The W3C Media Annotations Working Group (MAWG)[4] has the goal of improving the interoperability between multimedia metadata schemas. The proposed approach is to provide an interlingua ontology and an API designed to facilitate cross-community data integration of information related to media resources in the web, such as video, audio, and images. The set of core properties that constitute the Media Ontology 1.0 is based on a list of the most commonly used annotation properties from media metadata schemas currently in use. Like with MPEG Query Format, the aim of mapping is in this case the Media Ontology and mapping between a pair of formats is not supported.

Finally, there are several initiatives in the cultural heritage domain to provide services for mapping between library and museum metadata formats. The Vocabulary mapping framework[5] is a project connected with JISC[6]. It is an extension of the RDA/ONIX framework, which was developed in 2006 for libraries and the publishing industry as a framework for categorizing resources. It facilitates the transfer and use of resource description data.

The OCLC crosswalk service[7] is part of the Metadata Schema Transformation project, which tries to support library staff who often have to translate between different metadata standards. The project designed a self-contained crosswalk utility that can be called by any application that has to translate (supported) metadata records. The translation logic is executed by a dedicated XML application called the Semantic Equivalence Expression Language (SEEL), a language specification and a corresponding interpreter that is used to execute the transformations. The system is closed source and the publicly available information is really sparse.

The Athena mapping service[8] is an online service to convert in different XML formats (e.g. museum data) to the LIDO and to the Europeana format. Users can create a document collection on an Athena server by uploading single documents or sets of documents The service infers the schema from the presented documents. The definition of a mapping happens in two steps: (i) Identifying "items" in a tree view of a document and naming them, and (ii) define a mapping between the items and the target schema elements. Both steps are done by using drag and drop. Mapping possibilities include mapping structural elements, by repeating input elements, constant value assignment, a concatenate function and conditional mapping. When a mapping definition is finished, the

[4] http://www.w3.org/2008/WebVideo/Annotations
[5] http://cdlr.strath.ac.uk/VMF/background.htm
[6] http://www.jisc.ac.uk
[7] http://xwalkdemo.oclc.org/
[8] http://www.athenaeurope.org

transformation of the input documents can be initiated. The service currently supports concatenating values, but not splitting. Management and mapping of controlled vocabularies is supported.

3 Approach

Our approach uses a high-level intermediate representation of metadata elements. Metadata elements from a specific metadata format are described in relation to these generic elements. In addition, mapping templates on data type level are used. From these sources the code for a mapping problem between a pair of formats can be derived. We use an existing ontology, which has been extended for our purpose, to represent the formal semantics of the high-level intermediate representation. Since mapping instructions are derived from the ontology description, these mapping instructions are easier to maintain than hard coded mapping instructions. Furthermore the high-level intermediate representation serves as a hub for mapping between formats. Therefore hand-crafted one-to-one mappings between each pair of metadata formats are avoided, the mappings can be created automatically. Adding a new metadata format is done without side effects to existing definitions.

The core of this approach is the *meon* ontology[9] [8] which describes generic metadata elements and the relations between them. *meon* was originally developed to model metadata elements used throughout the audiovisual media production workflow in a format independent way in order to support content exchange and automation. Since concepts and relations coming from the *meon* ontology are reused and extended in our approach, the relations between metadata elements or groups of metadata elements used in this ontology are briefly introduced.

Definition relation. A metadata element (or group of elements) A defines another metadata element (or group of elements) B, if B can be derived without any semantic ambiguity from A by some mapping/conversion.

Equivalence relation. A metadata element (or group of elements) B is equivalent to another metadata element (or group of elements) B, if A defines B and B defines A.

Subtype relation. is used to describe that a metadata element is a more specific one of another metadata element.

In order the express these relations, the *meon* ontology, expressed in OWL-DL [3] models the classes `meon:Concept`, with subclasses `meon:AtomicConcept` and `meon:CompoundConcept` and three data type properties, namely `meon:contains`, `meon:defines` and `meon:subTypeOf`. The class `Concept` models the general concept represented by a metadata element in the ontology. Class `AtomicConcept` represents all single metadata elements, while class `CompoundConcept` describes groups of metadata elements containing at least two metadata elements. Property `contains` describes that an instance of class `AtomicConcept` is part of a

[9] Prefix `meon`: `http://www.20203dmedia.eu/meon#`

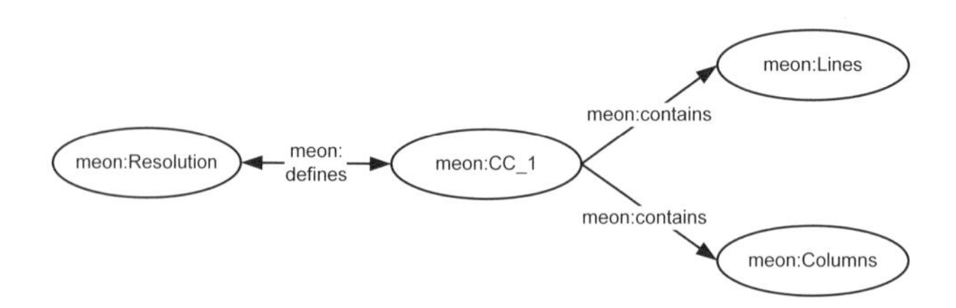

Fig. 1. *meon* example

group of metadata elements (i.e. part of an instance of class `CompoundConcept`). In order to describe that a metadata element (or group of elements) can be derived from another metadata element (or group of elements) by some mapping/conversion, the property `defines` is used. This property is used between instances of class `Concept`. Since the equivalence relation is just the result of a bidirectional definition relation, an appropriate use of this property is expressive enough to model also the equivalence relation. The property `subTypeOf` defines a relation between two `Concept` instances, of which one is a specialization of the other.

An example of the *meon* ontology for describing metadata elements and their relations between is shown in Figure 1. In this example *meon* is used to describe that the metadata element resolution is equivalent to the group of metadata elements containing lines and columns. Therefore resolution, lines, and columns are modeled as instances of class `AtomicConcept`.

`CC_1`, which is an instance of class `CompoundConcept`, represents the group of metadata elements containing lines and columns (denoted by the `contains` relation). To model the equivalence relation, two `defines` relations between `meon:Resolution` and `CC_1` are used.

The *meon* ontology has been extended to express mapping relations between metadata formats. In addition to the ontology of generic metadata concepts, specific ontologies are created for each format. They follow the same pattern as the *meon* ontology and include `meon:defines` relations between the format specific and the generic concepts. Figure 3 shows an example for formalizing the mapping relations of metadata elements expressed in MPEG-7 and Dublin Core via generic *meon* concepts, in this case `meon:creator`, `meon:producer`, `meon:contributor`, `meon:performer`, `meon:resolution`, `meon:lines` and `meon:columns`. It also models their relations, i.e. `meon:producer` is a subtype of `meon:creator`, `meon:performer` is a subtype of `meon:contributor`, and the compound of `meon:lines` and `meon:columns` is equivalent to `meon:resolution`. In the same manner the format specific concepts are defined. In MPEG-7, the creator element is used in combination with a role classification scheme to define types of creators and contributors. Therefore any specialized type of producer or contributor according to the used classification scheme is modeled as an subtype of class `mpeg7:Creator`. In our case, `mpeg7:Producer` and `mpeg7:Performer` are subtypes of `mpeg7:Creator`. In addition, another subtype `mpeg7:UnqualifiedCreator` is used to describe subtypes of `mpeg7:Creator` which are not explicitly modeled in the ontology. For describing the resolution in MPEG-7, `mpeg7:Width` and `mpeg7:Height` are added to the ontology. Furthermore the Dublin Core concepts `dc:Creator`, `dc:Contributor` and the DC Terms concept `dcterms:Extent` are added.

Definition relations are introduced to link these concepts to the *meon* concepts. `dc:Creator` is aligned with `meon:Creator` and `dc:Contributor` with `meon:Contributor`. Similarly, `mpeg7:Producer` is equivalent to `meon:Producer` and `mpeg7:Performer` is equivalent to `meon:Performer`. Furthermore there exists a equivalence relation between `mpeg7:UnqualifiedCreator` and `meon:Creator`, while there is only a definition relation between `meon:Contributor` and `mpeg7:UnqualifiedCreator`. In addition `mpeg7:Height` (resp. `mpeg7:Width`) is identical to `meon:Lines` (resp. `meon:Columns`) and `dcterms:Extent` is also equivalent to `meon:Resolution`.

Now it is possible to infer how concepts from different metadata formats are related by observing the relations among generic concepts and to the format specific concepts. In our example, it can be derived that `mpeg7:Producer` defines `dc:Creator` and `mpeg7:Performer` defines `dc:Contributor`. Since `dc:Creator` is the top-level term for all kind of creators, it can only define the `mpeg7:UnqualifiedCreator`. The same applies to `dc:Contributor`. This simple example thus also demonstrates typical limitations, that can only be overcome with additional, possible application specific information. Due to the lack of role qualification in Dublin Core, only a very coarse mapping to MPEG-7 is possible. If the classification scheme used by the MPEG-7 document is known, the situation can be improved. Then it is at least possible to decide whether an MPEG-7 role maps to creator or contributor, and in the other direction, a contributor could be mapped to a contributor term in the classification scheme, if that exists.

In order to retrieve mapping instructions between formats, format specific concepts are extended to carry additional information needed for mapping. Therefore the XPath location [1] and information about the data type representation of the metadata element are added to the ontology. For this purpose a new class `meon:DataTypeRepresentation` and new properties `meon:hasDataTypeRepresentation`, `meon:hasXPath` and `meon:hasDataTypeFormat` are introduced. This information is used to locate elements in the input and output documents and to apply the appropriate data type conversion. In Figure 4 the data type representations for the concepts `mpeg7:Producer` and `dc:Creator` are shown as an example. There can be several data type representations for a concept, typically associated with different XPath expressions. For example, the `mpeg7:Creator` element could contain the name of a person or organization, indicated by the value of its `mpeg7:Agent/@xsi:type` attribute.

We propose to use of XSL templates [6] for representing the data type conversion instructions. We define a table of such templates, which are annotated with their input and output data type formats. Once the inference is done, the input and output data types of the elements in the respective source and target formats are known and the appropriate template is found by looking up in the table. An example excerpt of such a table is shown in Table 1. Additionally, the

```
<xsl:template  name="Mpeg7_PersonType__TaggedText">
  <xsl:param name="tag">xxx</xsl:param>
  <xsl:for-each select="mpeg7:Name">
    <xsl:element name="{$tag}">
      <xsl:call-template name="Mpeg7_Name__SimpleText"/>
    </xsl:element>
  </xsl:for-each>
</xsl:template>

<xsl:template name="Mpeg7_Name__SimpleText">
  <xsl:for-each select="mpeg7:GivenName">
    <xsl:value-of select="."/>
    <xsl:text> </xsl:text>
  </xsl:for-each>
  <xsl:for-each select="mpeg7:FamilyName">
    <xsl:value-of select="."/>
    <xsl:text> </xsl:text>
  </xsl:for-each>
</xsl:template>
```

Fig. 2. Example of XSL template for transforming the name of a structured MPEG-7 person type into a string

Table 1. Example excerpt of data type template table

XSLT	Input data type format	Output data type format
T1	XML_Structured_GivenName_FamilyName	String_FirstName_LastName
T2	XML_Structured_Organization_Name	String_Organization_Name
T3	XML_Attributes_Width_Height	String_Width_Height

required XSL templates to transform the name of a structured MPEG-7 person type into a simple string are depicted in Figure 2.

In our example, it can be derived that `dcterms:Extent` defines both `mpeg7:Height` and `mpeg7:Width`, and also that the compound of `mpeg7:Height` and `mpeg7:Width` defines `dcterms:Extent`. However, we need to apply a single template (T3 in Table 1) that concatenates the values of the width and height attributes in the MPEG-7 documents to the string value of `dcterms:Extent`. This template is associated with both the `mpeg7:Width` and `mpeg7:Height` concepts. The inference process yields a new compound concept containing these two concepts which defines the desired output (`dcterms:Extent`) via `meon:Resolution` (cf. Figure 3). We then look for an appropriate data type template that is associated with all the concepts contained in the inferred compound concept. This data type template is applied to convert the group of source elements (`mpeg7:Height` and `mpeg7:Width`) to the target element (`dcterms:Extent`).

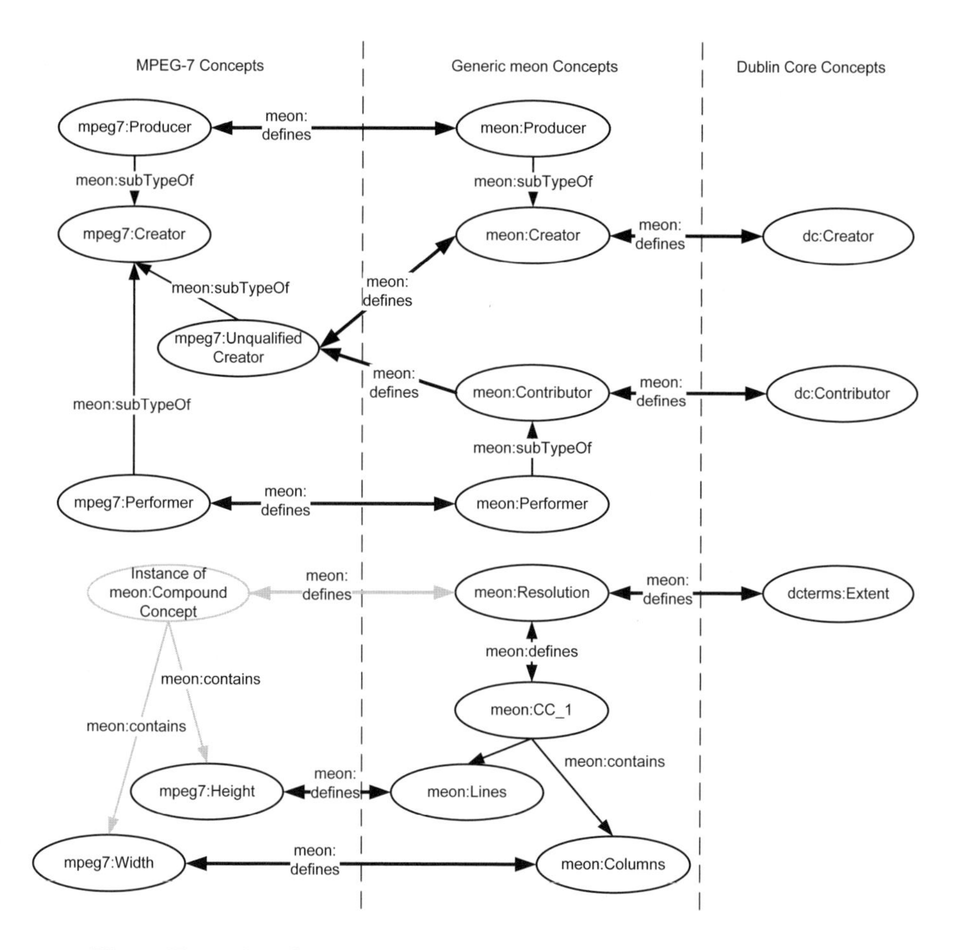

Fig. 3. Examples of mapping elements between MPEG-7 and Dublin Core

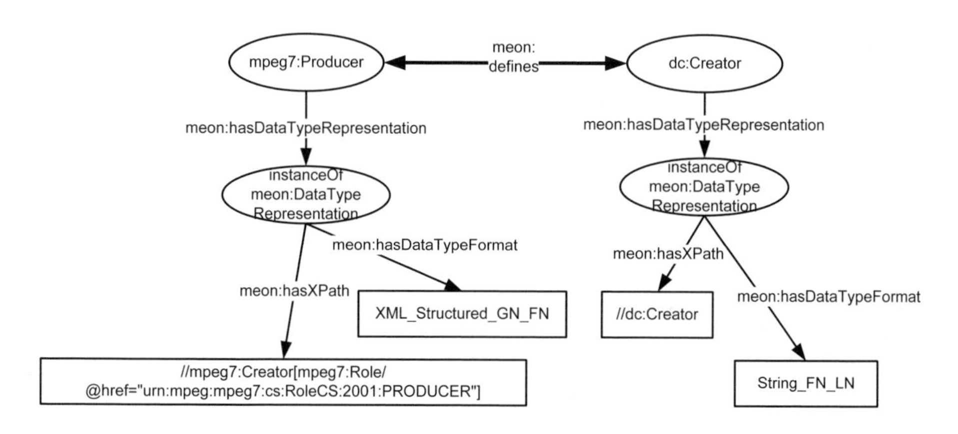

Fig. 4. Data type representation in the ontology

```
...
<Creator>
  <Role href="urn:mpeg:mpeg7:cs:RoleCS:2001:PERFORMER">
    <Name>Performer</Name>
  </Role>
  <Agent xsi:type="PersonType">
    <Name>
      <GivenName>Mick</GivenName>
      <FamilyName>Jagger</FamilyName>
    </Name>
  </Agent>
</Creator>
...
<Creator>
  <Role href="urn:mpeg:mpeg7:cs:RoleCS:2001:PRODUCER">
    <Name>Producer</Name>
  </Role>
  <Agent xsi:type="OrganizationType">
    <Name>Acme Inc.</Name>
  </Agent>
</Creator>
...
<Creator>
  <Role href="urn:mpeg:mpeg7:cs:RoleCS:2001:">
    <Name>Creator</Name>
  </Role>
  <Agent xsi:type="PersonType">
    <Name>
      <GivenName>Yoshiaki</GivenName>
      <FamilyName>Shibata</FamilyName>
    </Name>
  </Agent>
</Creator>
...
<VisualCoding>
  <Frame height="640" width="480"/>
</VisualCoding>
```

Fig. 5. Fragments from an MPEG-7 example document

```
...
<dc:contributor>Mick Jagger</dc:contributor>
<dc:creator>Acme Inc.</dc:creator>
<dc:creator>Yoshiaki Shibata</dc:creator>
<dcterms:Extent>640x480 px</dcterms:Extent>
...
```

Fig. 6. Dublin Core example fragment

Fig. 7. Screenshot of the metadata converter prototype application

4 Prototype Application

Following the mapping approach described above, the current prototype implementation of the mapping service is based on an ontology describing how metadata concepts coming from different metadata formats are related. This formalization is used to infer concrete mapping instructions between a pair of metadata formats. While the ontology is expressed in OWL-DL, the mapping instructions are encoded as XSL transformations [6] in order to perform the mapping of an instance document. The workflow of the current prototype implementation is as follows. The metadata concepts and mapping relations in the ontology are annotated with additional information needed for mapping, i.e. format specific metadata with the corresponding XPath information and information on data type representation. Using reasoning techniques over the ontology it is now possible to infer which data type templates are needed for the mapping between different format specific concepts. A SPARQL[10] query is used to retrieve all necessary mapping information from the ontology. Then the next step is to generate a XSL template out of the mapping information. The

[10] http://www.w3.org/TR/rdf-sparql-query/

XSL template document with mapping instructions for all input elements can be applied to a input document instance.

Figures 5 and 6 show the MPEG-7 and Dublin Core XML fragments corresponding to the example described above. As we have discussed earlier, mapping of the MPEG-7 fragment in Figure 5 yields the Dublin Core fragment in Figure 6, but not vice versa. Without further knowledge, both `dc:Creator` and `dc:Contributor` would map to `mpeg7:UnqualifiedCreator`.

The metadata semantic converter application (Figure 7) allows the user to choose an example data set in MPEG-7 XML format and to convert it to Dublin Core XML format. The application displays the conversion steps together with the XML data involved. The application is written in Java using the GWT[11]. The conversion steps are deployed as asynchronous web services on Apache Tomcat 6.0[12]. The server application uses the Jena[13] library for reading and writing RDF and OWL files. The XSLT transformation is done using the Xalan[14] library. The prototype application is available at `http://prestoprime.joanneum.at/`.

5 Conclusion and Future Work

We have presented an approach for mapping between multimedia metadata formats using an ontology, that represents high-level format independent metadata concepts, as well as format specific concepts and their relations to the high-level concepts, together with data type templates. By using an intermediate format independent representation, the approach attempts to avoid pair-wise mappings between formats, thus breaking the need for a quadratic number of mappings. We have implemented mapping between MPEG-7 and Dublin Core in a prototype application. This example demonstrates the feasibility of the approach but also shows limitations due to limited expressiveness of formats.

The proposed approach can support complete mappings between two metadata formats as well as partial mappings in case of formats that do not sufficiently overlap. Partial mappings occur in cases where the *meon* ontology does not include appropriate concepts for format specific properties or mappings are incomplete due to limitations of the expressiveness of the formats involved. This also means that adding new formats does not require changing the *meon* ontology, but only adding concepts and relations if necessary. In cases where a property is specific to a format, there is no need to add it to *meon*, as it will not be possible to map it to any other format. In cases where a property is more general, it can be added to *meon*, and relation with existing concepts can be defined. Note that this does not require any modifications in the format specific ontologies for exisiting formats nor in exiting mapping templates.

The definition of the data types formats via long unique names is a temporary solution. In the next version, we will integrate the description of the data types

[11] Google Web Toolkit, `http://code.google.com/webtoolkit/overview.html`
[12] `http://tomcat.apache.org/index.html`
[13] `http://jena.sourceforge.net/`
[14] `http://xml.apache.org/xalan-j/`

and their related properties (e.g., the specifics of a person name string) in the ontology to have this information available for machine processing.

It has been discussed in Section 1.2 that mappings between formats with different expressivity or incongruent semantics can be ambiguous or incomplete. In a practical application some of these limitations can be overcome by defining rules based on additional context knowledge, such as about vocabularies (e.g., MPEG-7 classification schemes) used in the documents used as values or qualifiers of elements, or knowledge about the specific flavour of a format used. Such rules can help to reduce the potential information loss in the mapping process. In order to support users in the definition of such rules, a tool for defining theses rules will be necessary.

One further issue that needs to be addressed in future works are mappings between formats that differ w.r.t. the structure of the description, e.g. mapping from a flat to a hierarchical scheme. Like with incongruent semantics, there might be cases where such issues can only be resolved unambiguously with addition application specific knowledge, especially when mapping from simpler to more complex structures.

Acknowledgments. The research leading to this paper was partially supported by the European Commission under the ICT contract FP7 231161, "Presto-PRIME" (`http://www.prestoprime.eu`).

References

1. Clark, J., DeRose, S. (eds.): XML Path Language (XPath) Version 1.0. W3C Recommendation (1999), `http://www.w3.org/TR/xpath/`
2. Cruz, I., Xiao, H., Hsu, F.: An Ontology-Based Framework for XML Semantic Integration. In: International Database Engineering and Applications Symposium, Coimbra, Portugal, pp. 217–226 (2004)
3. Dean, M., Schreiber, G.: OWL Web Ontology Language: Reference. W3C Recommendation, February 10 (2004), `http://www.w3.org/TR/owl-ref/`
4. Döller, M., Stegmaier, F., Kosch, H., Tous, R., Delgado, J.: Standardized Interoperable Image Retrieval. In: ACM Symposium on Applied Computing (SAC), Track on Advances in Spatial and Image-based Information Systems, ASIIS (2010)
5. Döller, M., Tous, R., Gruhne, M., Yoon, K., Sano, M., Burnett, I.: The MPEG Query Format: On the way to unify the access to Multimedia Retrieval Systems. IEEE Multimedia 15(4), 82–95 (2008)
6. Kay, M. (ed.): XSL Transformations (XSLT) Version 2.0. W3C Recommendation (2007), `http://www.w3.org/TR/xslt20/`
7. ETSI. ETSI TS 102 822-3-1 V1.5.1: Broadcast and On-line Services: Search, select, and rightful use of content on personal storage systems ("TV-Anytime") - Part 3: Metadata (May 2009), `http://www.etsi.org`
8. Höffernig, M., Bailer, W.: Formal Metadata Semantics for Interoperability in the Audiovisual Media Production Process. In: Workshop on Semantic Multimedia Database Technologies (SeMuDaTe 2009), Co-located with the 4th International Conference on Semantic and Digital Media Technologies (SAMT 2009). CEUR-WS, vol. 539 (2009)

9. Kondylakis, H., Doerr, M., Plexousakis, D.: Mapping language for information integration. Technical Report TR385, Institute of Computer Science, FORTH-ICS (2006)
10. MPEG-7. Multimedia Content Description Interface. ISO/IEC 15938 (2001)
11. Poppe, C., Martens, G., Mannens, E., Van de Walle, R.: Personal Content Management System: A Semantic Approach. Journal of Visual Communication and Image Representation 20(2), 131–144 (2009)
12. Xing, G., Xia, Z., Ernest, A.: Building automatic mapping between XML documents using approximate tree matching. In: ACM Symposium on Applied Computing (SAC), Seoul, Korea, pp. 525–526 (2007)
13. Yang, X., Lee, M., Ling, T.W., Tok, L., Ling, W.: Resolving Structural Conflicts in the Integration of XML Schemas: A Semantic Approach. In: Song, I.-Y., Liddle, S.W., Ling, T.-W., Scheuermann, P. (eds.) ER 2003. LNCS, vol. 2813, pp. 520–533. Springer, Heidelberg (2003)

Combining Visual and Textual Modalities for Multimedia Ontology Matching

Nicolas James, Konstantin Todorov, and Céline Hudelot

MAS Laboratory, École Centrale Paris, F-92 295 Châtenay-Malabry, France
{nicolas.james,konstantin.todorov,celine.hudelot}@ecp.fr

Abstract. Multimedia search and retrieval are considerably improved by providing explicit meaning to visual content by the help of ontologies. Several multimedia ontologies have been proposed recently as suitable knowledge models to narrow the well known semantic gap and to enable the semantic interpretation of images. Since these ontologies have been created in different application contexts, establishing links between them, a task known as ontology matching, promises to fully unlock their potential in support of multimedia search and retrieval. This paper proposes and compares empirically two extensional ontology matching techniques applied to an important semantic image retrieval issue: automatically associating common-sense knowledge to multimedia concepts. First, we extend a previously introduced matching approach to use both textual and visual knowledge. In addition, a novel matching technique based on a multimodal graph is proposed. We argue that the textual and visual modalities have to be seen as complementary rather than as exclusive means to improve the efficiency of the application of an ontology matching procedure in the multimedia domain. An experimental evaluation is included.

1 Introduction

In recent years, many research efforts have been directed towards the problem of improving search and retrieval in large image collections by providing semantic annotations in a fully automatic manner. Ideally, semantic image annotation results in a linguistic description of an image, which, in the current state of affairs, is often only related to perceptual manifestations of semantics. Indeed, most of the existing approaches are based on the automatic detection of semantic concepts from low level features with machine learning techniques. Nevertheless, as explained in [10], the image semantics cannot be considered as being included explicitly and exclusively in the image itself. It rather depends on prior knowledge and on the context of use of the visual information. In consequence, explicit formal knowledge bodies (ontologies) have been growingly used to relate semantics and images. Their application in the multimedia domain aims at improving image search and retrieval by providing high-level semantics to visual content, thus facilitating the interface between human and artificial agents and narrowing the well-known *semantic gap* between *low-level visual features* and *high-level meaning* [19].

T. Declerck et al. (Eds.): SAMT 2010, LNCS 6725, pp. 95–110, 2011.

However, differences in the scopes and purposes of these ontologies (reviewed in Section 2) as well as in their application contexts tend to result in various heterogeneities on terminological, conceptual and / or semantic level. Therefore, relating these knowledge resources, a process termed as *ontology matching* [5], is crucial in order to fully unlock their potential in support of multimedia search and retrieval - a field in which ontology matching has found little application in contrast to its use in the semantic web domain. To accomplish an ontology matching task one could rely on the instances contained in the ontologies concepts (*extensional* or *instance-based* matching), make use of the relations that hold between the different concepts (*structural* matching), measure the similarities of the concept names and their lexical definitions (*terminological* matching), etc. In the case of multimedia ontologies, which often come equipped with sets of annotated images, extensional matching is a suitable paradigm since it enables to benefit from both the visual and the textual knowledge.

This paper considers two generic instance-based ontology matching techniques - one based on variable selection (developed in a previous study of the same authors [12], [23]) and another, novel approach, exploring the benefits of discovering correlations in a multimodal graph. We apply and compare these approaches in the context of an important semantic image retrieval problem: associating common-sense knowledge to multimedia concepts. In particular, the paper proposes to narrow the semantic gap by matching a common sense ontology (WordNet [15] associated to the image database LabelMe[18]) with a specific multimedia ontology (LSCOM [20] associated to the TRECVID2005 development data set).

Since our matching approaches rely on extensional information, it is important to explore and make use of all possible instance-based knowledge that can be made available. In extensional terms, these two resources can be considered as bi-modal, each possessing a *visual* and a *textual* modality. On one hand, the concepts of these ontologies serve to annotate a given set of images which can be considered as instances of these concepts. On the other hand, every image can be assigned a text document by taking the concepts that it is annotated by and their corresponding textual definitions (LSCOM definitions or WordNet glosses). In order to apply the suggested matching approaches, one can rely on either of the two modalities and we will refer to the two resulting types of matching as, respectively, *visual matching* and *textual matching*. What the paper investigates more closely, are the benefits of using both in combination, instead of each of them in isolation. The variable selection model is able to work with only one modality at a time and an integration of the results have to be performed *post factum*. Since we are primarily interested in obtaining concept correspondences based on the visual characteristics of the images in the two datasets, we will rely on the visual modality to produce a baseline matching, which will be later adjusted and refined by the help of the textual modality. A potential advantage of the graph-based model is that it allows the simultaneous, *built-in* application of the two modalities in the matching process.

The rest of the article is structured as follows. Next section reviews related work. Section 3 provides a summary of the generic instance-based ontology matching techniques that we use. The application of these methods on visual and textual instances is described in detail in Section 4: we report experimental results of these matchings and discuss the benefits of their integrated interpretation. We conclude in Section 5.

2 Related Work

Despite many recent efforts to provide approaches for automatic annotation of images with high-level concepts [11], the semantic gap problem is still an issue for the understanding of the *meaning* of multimedia documents. In this context, many knowledge models or ontologies have been proposed to improve multimedia retrieval and interpretation by the explicit modeling of the different relationships between semantic concepts.

In particular, many generic large scale multimedia ontologies or multimedia concept lexicons together with image collections have been proposed to provide an effective representation and interpretation of multimedia concepts [21,20,3]. We propose to classify these ontologies in four major groups: (1) semantic web multimedia ontologies often based on MPEG-7 (a review can be found in [3]) (2) visual concept hierarchies (or networks) inferred from inter-concept visual similarity contexts (among which VCNet based on Flickr Distance [26] and the Topic Network of Fan [6]), (3) specific multimedia lexicons often composed of a hierarchy of semantic concepts with associated visual concept detectors used to describe and to detect automatically the semantic concept of multimedia documents (LSCOM [20], multimedia thesauri [22], [21]) and (4) generic ontologies based on existing semantic concept hierarchies such as WordNet and populated with annotated images or multimedia documents (ImageNet [4], LabelMe [18]).

The reasoning power of ontological models has also been used for semantic image interpretation. In [2],[9] and [17], formal models of application domain knowledge are used, through fuzzy description logics, to help and to guide semantic image analysis. Prior knowledge on structured visual knowledge represented by an And-or graph (stochastic grammars) has been proved to be very useful in the context of image parsing or scene recognition in images [28]. While these different formal models are highly integrated in multimedia processing, their main drawback is that they are specific to the application domain.

All these ontologies have proved to be very useful mainly in the context of semantic concept detection and automatic multimedia annotation but many problems still remain among which the interoperability issue between visual concepts and high level concepts. To solve this issue, some ontology-based infrastructures have also been proposed to guide image annotation [1]. These infrastructures are mainly based on different ontologies (multimedia ontologies, application domain ontologies and top ontologies for interoperability purposes) and the link between the different ontologies is often done manually. In [21], the authors also propose to build a multimedia thesaurus by linking manually 101 multimedia concepts with WordNet synsets.

Due to the fact that these large scale multimedia ontologies are often dedicated to (or initially built for) particular needs or a particular application, they often tend to exhibit a certain heterogeneity which allows their use as complementary knowledge sources. For instance LSCOM was built for video news annotation purposes, while the scope of WordNet/LabelMe is rather general and common-sense. Hence, these ontologies differ both in their conceptual content (number, granularity and genericity of the concepts) and in their usage (LSCOM is dedicated to multimedia annotation and therefore the extensional and terminological knowledge that it assigns to each concept is defined by the visual appearances of this concept). While studies have been done to analyze the different inter-ontology concept similarities in different multimedia ontologies [13], to the best of our knowledge, there are no approaches in the state of the art which propose a cross analysis and a joint use of these different and complementary resources.

3 Instance-Based Ontology Matching

We propose a methodology to narrow the semantic gap by matching two complementary resources: a *visual* ontology and a *semantic* thesaurus. Contrarily to [21], we suggest to accomplish this matching in an automatic manner. We apply a generic extensional ontology matching approach based on discovering cross-ontology concept similarities via variable selection, which has been previously introduced for matching textually populated ontologies [23]. In [12], we propose a first extension of this approach based on visual extensional knowledge. In the framework of this paper, the approach has been extended to use both *textual* and *visual* knowledge with the objective to combine both in the concept alignment process. In addition, we suggest a novel matching technique, based on a multi-model graph and a Random Walk with Restart (RWR). In the sequel, we will describe the main elements of these approaches in a generic manner, by referring to an abstract notion of *instance*, without specifying whether it comprises a text or a multimedia document and how precisely it is represented. We only assume that each instance is indeed representable as a real-valued vector. We start by giving several assumptions and definitions.

An ontology is based on a set of *concepts* and *relations* defined on these concepts, which describe in an explicit and formal manner the knowledge about a given domain of interest. In this paper, we are particularly interested in ontologies, whose concepts come equipped with a set of associated instances, referred to as **populated ontologies** and defined as tuples of the kind $O = \{C, \texttt{is\_a}, R, I, g\}$, where C is a set whose elements are called concepts, $\texttt{is\_a}$ is a partial order on C, R is a set of other relations holding between the concepts from the set C, I is a set whose elements are called instances and $g : C \to 2^I$ is an injection from the set of concepts to the set of subsets of I. In this way, a concept is *intensionally* modeled by its relations to other concepts, and *extensionally* by a set of instances assigned to it via the mapping g. By assumption, every instance can be represented as an n dimensional real-valued vector, defined by n input **variables** of some kind (the same for all the instances in I).

To build a procedure for ontology matching, we need to be able to measure the **pair-wise similarity of concepts**. The measures used in the current study are based on *variable selection* (Section 3.1) and on correlations discovered by a random walk in a *mixed multimedia graph* (Section 3.2).

3.1 Variable Selection-Based Method (VSBM)

Variable selection [7] is defined as a procedure for assigning *ranks* to the input variables with respect to their importance for the output, a ranking criterion provided. On these bases, we propose to evaluate concept similarity by comparing the ranks assigned to the input variables w.r.t. two given concepts.

We define a binary training set S_O^c for each concept c from an ontology O by taking I, the entire set of instances assigned to O and labeling all instances from the set $g(c)$ as *positive* and all the rest $(I \backslash g(c))$ as *negative*. By the help of a variable selection procedure performed on S_O^c (i.e. evaluating the importance of the input variables w.r.t. the concept c), we obtain a representation of the concept c as a list

$$L(c) = (r_1^c, r_2^c, ..., r_n^c), \tag{1}$$

where r_i^c is the rank associated to the ith variable. To compute a rank per variable and per concept, we apply a standard *Point-wise Mutual Information* criterion approximated for a variable v_i and a concept c by

$$r_i^c = PMI(v_i, c) = \log \frac{A \times |I|}{(A + C) \times (A + B)}, \tag{2}$$

where A is the number of co-occurrences of v_i and c, B is the number of non-zero occurrences of v_i out of c, C is the number of zero occurrences of v_i within c and $|\cdot|$ stands for set cardinality [27].

Given two source ontologies O and O', a representation as the one in (1) is made available by following the described procedure for every concept of each of these ontologies. The similarity of two concepts, $c \in O$ and $c' \in O'$ is then measured in terms of their corresponding representations $L(c)$ and $L(c')$. Several choices of a similarity measure based on these representations are proposed and compared in [23]. In the experimental work contained in this paper, we have used Spearman's measure of correlation and the n'-TF similarity measure. Spearman's coefficient is given by

$$s_{Spear}(c, c') = 1 - 6 \frac{\sum_i (r_i^c - r_i^{c'})^2}{n(n^2 - 1)}. \tag{3}$$

The n'-TF (n' Top Features) simply measures the size of the intersection of the subsets of the $n' < n$ top variables (i.e. the ones with highest ranks) according to the lists $L^{vars}(c)$ and $L^{vars}(c')$:

$$s_{n'TF}(c, c') = \frac{|\{v_{i_1}, ..., v_{i_{n'}}\} \cap \{v_{j_1}, ..., v_{j_{n'}}\}|}{n'}, \tag{4}$$

where $v^c_{i_p}$ stands for the variable which has a rank $r^c_{i_p}$ and $v^c_{j_q}$ - for the variable which has a rank $r^{c'}_{j_q}$

In the sequel, we will be interested in applying the measures above in order to represent a concept c from ontology O by a list of pairs of the kind (s_i, c'_i), where s_i (a shortcut for $s_i(c, c'_i)$) is the similarity score (issued from either measure (3) or (4)) of the concept c and the concept $c'_i \in O'$, $i = 1, ..., k$ with k the cardinality of the concept set of O'. We will denote the list of such pairs corresponding to the concept $c \in O$ by

$$L^{sim}(c) = \{(s_1, c'_1), ..., (s_k, c'_k)\}. \tag{5}$$

Provided the choice of a threshold $k' \leq k$, we will define for each concept $c \in O$ the *matching* $L^{sim}_{k'}(c)$ by keeping only those k' concepts from O' which have the highest similarity scores with respect to c. An *alignment* of O to O' will be defined as the set of matchings $A(O, O') = \{L^{sim}_{k'}(c_j)\}^l_{j=1}$, where l is the cardinality of the concept set of O.

3.2 Graph-Based Method (GBM)

Graph-based procedures are well-known approaches for evaluating the similarity between objects, like our concepts. These approaches have been used in several domains: ranking algorithms for information retrieval [8], automatic image annotation [16], [25], data analysis and word sense disambiguation [14]. The idea is to exploit the relationships between objects and the different aspects of these objects. In our instance-based ontology matching framework, we have objects of different kinds: (1) concepts, (2) concept instances (i.e. images), and (3) features relevant to the images. We use a method based on the Mixed-Multimedia Graph (*MMG*) and the Random Walks with Restarts algorithm proposed in [16]. Fig. 1 represents a special case of a MMG in the scope of our concept matching procedure.

The MMG graph is well adapted for multimedia document processing because it allows to mix heterogeneous kinds of information, like illustrated in Fig. 1. For each instance we have: (1) a concept node, (2) a textual representation of the instance and (3) a visual representation of the instance. For our experiments, the textual representation is based on a bag-of-words model built from the textual definition of all the concepts associated to an instance (the instances are multi-annotated), and the visual representation is based on histograms of visual words computed over the instances. The kind of the instance representation is seen as a modality, the graph is modular and a modality can be easily used or not. As we will see in the experiments (Section 4.1), we have used both the uni-modal and the bi-modal versions of the graph. Regardless to the chosen modality, the graph is completed with Nearest Neighbor (NN) links between the nodes of each modality. The similarity function needed to compute these links depends on the modality type, therefore we need a function for computing the similarity between two textual representations and another function for computing the similarity between two visual representations.

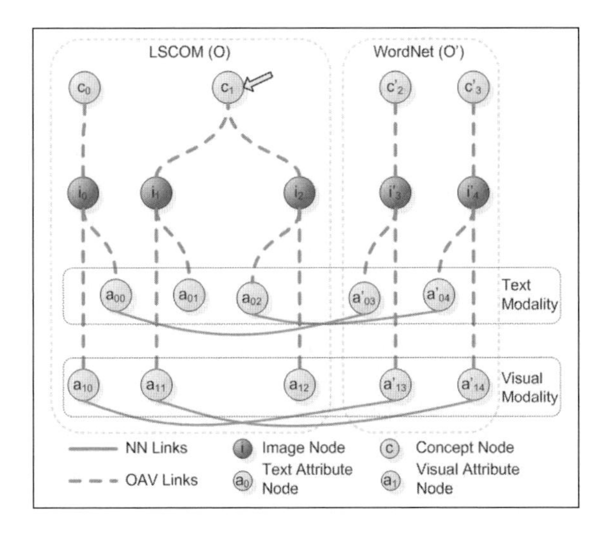

Fig. 1. An MMG Graph

The process of discovering concept similarities consists in finding correlations between a specific concept of the ontology O and the concepts of the ontology O'. We use the RWR algorithm as described in [16]. The random walk starts at a fixed concept node - the one for which we search similar concepts in the ontology O' (see the concept c_1 in Fig. 1). At each step, the walker can either choose a link in the set of associated links to the node on which it is, or go to its starting point with a probability p (experimentally set to 0.8 as in [16]). A precise description with implementation details of the algorithm can be found in [24].

The probability that the walker is at node c', called the *steady state probability*, $\mu_{c_1}(c')$, can be interpreted like an affinity measure between the node c_1 and c'. Therefore, if we consider the results only for concept nodes, a high similarity between $c_1 \in O$, and $c' \in O'$ is observed when the probability $\mu_{c_1}(c')$ is high.

4 Aligning Two Multimedia Resources

The ontology matching techniques described above can be applied for any two ontologies whose concepts are used to label a set of real-world instances of some kind. Based on these techniques, we will align two complementary multimedia knowledge resources by using and integrating the visual and textual modalities of their extensions.

We chose, on one hand, LSCOM [20] – an ontology dedicated to multimedia annotation. It was initially built in the framework of TRECVID[1] with the criteria of concept usefulness, concept observability and feasibility of concept automatic detection. LSCOM is populated with the development set of TRECVID 2005 videos. On the other hand, we used WordNet [15] populated with the LabelMe dataset [18].

[1] http://www-nlpir.nist.gov/projects/tv2005/

4.1 Experimental Setting

In our experimental work, we have used a part of the LSCOM ontology, LSCOM_Annotation_v1.0[2], which is a subset of 449 concepts from the initial LSCOM ontology, and is used for annotating 61,517 images from the TRECVID2005 development set. Since this set contains images from broadcast news videos, the chosen LSCOM subpart is particularly adapted to annotate this kind of content, thus contains abstract and specific concepts (e.g. SCIENCE_TECHNOLOGY, INTERVIEW_ON_LOCATION). To the contrary, our subontology defined from WordNet populated with LabelMe (3676 concepts) is very general considering the nature of LabelMe, which is composed of photographs from the daily life and contains concepts such as CAR, COMPUTER, PERSON, etc.

To provide a low-scale evaluation of the suggested approaches, we chose five concepts from the LSCOM ontology (BUS, COMPUTER, PEDESTRIAN_ZONE, SPEAKER_AT_PODIUM, SPORT) populated with 2317 images, and thirteen concepts from the WordNet ontology (ARM, CAR, GRASS, HEAD, LEG, PERSON, PLANT, PLATE, ROAD, SIDEWALK, TORSO, TREE, WHEEL) populated with 4964 images. The choice of the selected concepts was made on the basis of several criteria: (1) the number of associated instances, (2) the lack of semantic ambiguity in our dataset for every selected concept, (3) for WordNet only: a high confidence (arbitrarily decided) in the discrimination of the concept using only perceptual information, (4) the presence of contextually bound cross-ontology concepts (such as BUS and CAR) as well as contextually isolated concepts (i.e. dissimilar to all the other concepts such as PLATE).

We draw the readers attention to the fact that the similarities of the concepts should be interpreted strictly within the extensional nature of their definitions and not in terms of any possible intuitive or common sense definition. Our methods imply that two concepts are similar if their corresponding instances contain similar visual or textual characteristics (the instances of two similar concepts contain some identical objects). In some cases, these similarity scores are in agreement with the common sense, but they are not in other cases. In that line of thought, taking two concepts (one from each of the ontologies) with identical names (e.g. BUS in WordNet and BUS in LSCOM) is not relevant for evaluating the quality of the alignments.

In the remainder of the section, we will first present and discuss results from the visual and textual matchings of the selected sets of concepts separately. We will further propose a method to integrate the two matching types. As matching procedures we have used the VSBM method with two different concept similarity measures - Spearman's correlation and the n'-TF measure[3]. Additionally, we have tested the GBM approach by using either only the visual or only the textual modality and by using both modalities simultaneously. This results in three independent alignments per matching type which, to improve readability, are all gathered at the end of the paper.

---

[2] http://www.ee.columbia.edu/ln/dvmm/lscom/

[3] Pearson's measure, also discussed in [23] showed to compete closely with Spearman's.

4.2 Visual Matching

Instances and Representation. To construct image features, we use a bag-of-words model with a visual codebook of size 900, built classically using the well known SIFT descriptor and a K-Means algorithm. The quantification of the extracted SIFT features was done over all the instances associated to the selected concepts (from both LSCOM and WordNet) by using only the distinct objects in each image instead of the entire image in order to extract the SIFT features. The variables which describe an image are then the bins of the histograms of codewords corresponding to this image.

Results and Discussion. The results of matching the 5 LSCOM concepts against the 13 WordNet concepts by following the variable selection-based matching procedure described above are presented in Tables 1 and 2 and the results from the GBM method are shown in Table 3. As introduced in (5), we provide for every LSCOM concept (in the top row) a list of pairs (score, WordNet concept) in a descending order of their importance with respect to this concept. The scores in Table 1 and Table 2 are issued from applying the similarity measures (3) and (4), respectively, whereas the scores in Table 3 are correlations found by the help of the graph-based method.

As a general tendency, the WordNet concepts PERSON and HEAD tend to appear up in the lists, whereas the concept PLATE achieves mostly low scores. These results are coherent with the nature of our data, since the concept PLATE stands alone in our selection of concepts, whereas the concepts PERSON and HEAD are highly relevant for the TRECVID dataset, containing shots from news videos where often we have a presenter or a speaker. For Table 3, some remarks about the graph construction have to be taken into account. The similarity used to compute the nearest-neighbor links is a Minkowski distance. Due to the nature of the LSCOM/TRECVID data (the images are visually very close to one another within TRECVID), we have taken into account only NN links from ontology O to O' (or vice-versa) in order to get a well connected graph. Without this constraint, we would have a graph with two disconnected components. However, this explains the results in Table 3 where the top 5 concepts are the same in all lists (subject to a permutation).

We observe examples of a lack of coherence between intuitive interpretations and achieved matchings as discussed previously in this section. For example, SIDEWALK w.r.t. COMPUTER is intuitively an erroneous matching in contrast to LEG and PERSON w.r.t. SPEAKER_AT_PODIUM which is intuitively coherent (Table 1 and 2).We note that this is a perceptually induced conceptual mismatch, i.e. a bias, which is due to co-occurrences of *visual* objects within the instances of both concepts. In our example, images of SIDEWALK tend to contain the object *person*, so do images of COMPUTER, although COMPUTER and SIDEWALK are unrelated. In order to account for this problem, we suggest that a post-processing of the obtained matchings has to be performed with the objective to re-rank the WordNet concepts w.r.t. their importance for the respective LSCOM concept. To these ends, we perform a textual matching with the objective to complement the results achieved by the visual matching and filter out undesired alignments.

4.3 Textual Matching

Instances and Representation. We present the results of the matching of the two selected sets of concepts, by using this time as instances textual documents, relevant to these concepts. A text document has been generated for every image, by taking the names of all concepts that an image contains in its annotation, as well as the (textual) definitions of these concepts (the LSCOM definitions for TRECVID images or the WordNet glosses for LabelMe images). An example is shown in Fig. 2. After a phase of standard text-processing (lemmatization and stop-word filtering), a vocabulary of size 544 has been constructed for the corpus containing the documents generated as instances for the two ontologies. Every textual instance is represented as a *tf-idf* vector of dimension 544.

107_Standing One or more people standing up. 227_Bus Shots of a bus. 224_Outdoor Shots of Outdoor locations. 217_Person Shots depicting a person. The face may be partially visible. 202_Crowd Shots depicting a crowd. 181_Adult Shots showing a person over the age of 18. 104_Male_Person One or more male persons. 290_Daytime_Outdoor shots that take place outdoors during the day. 316_Group We defined a group as 3-10 people. 109_Windows An opening in the wall or roof of a building or vehicle fitted with glass or other transparent material.

Fig. 2. The LSCOM concept BUS: a visual and a textual instance

Results and Discussion. To derive the textual similarity scores, we have applied the same procedures as those applied for the visual matching. For the VSBM matching, we first scored the variables by the help of a mutual information-based variable selection and then measured concept similarities by the help of Spearman's measure of correlation and n'-TF (equations (3) and (4)). We note that in this case the variables that define our instances are actual (lemmatized) terms that appear in the corpus with a certain (sufficiently high) frequency. The results of these matchings are presented in an analogous manner to the visual matching in Tables 4 and 5. The similarity scores achieved by applying the graph-based approach on our textual data are found in Table 6.

We observe that, while some of the correlations already found by the visual matching are confirmed (e.g. the low scores of the stand-alone concept PLATE), some of the WordNet concepts achieve different scores through the textual matching (e.g. the problematic SIDEWALK). This confirms the initial hypothesis that the two matching modalities are complementary and neither is to be applied self-dependently. The integration and the proper interpretation of the results of the two matching types is the subject of the following sub-section.

4.4 Integration of the Textual and the Visual Modalities

***A posteriori* pruning with VSBM** As a result of the matchings, every WordNet concept is assigned two scores w.r.t. each LSCOM concept - a visual similarity score s_v^i and a textual similarity score s_t^i.

In order to identify problematic matchings (e.g. (COMPUTER, SIDEWALK)), we propose an algorithm which serves to prune the list of most important (w.r.t. a given LSCOM concept) WordNet concepts. We compute for every LSCOM concept the quantities $s_{\delta}^{i} = |s_{t}^{i} - s_{v}^{i}|$, $\forall i = 1, ..., k$, (k is here the number of WordNet concepts) represented in Tables 7 and 8. The scores s_{t}^{i} and s_{v}^{i} are integers corresponding to the (real) similarity scores. When multiple consecutive concepts achieve identical scores (a likely case when applying the n'-TF measure) the same rank is attributed to each of these concepts. We take as a basis the matching achieved by using the visual modality and we fix a number k' of concepts to be kept. Our algorithm relies on the heuristics that the WordNet concepts c_{i}' for which the corresponding s_{δ}^{i} is too large (w.r.t. an experimentally set threshold) should be identified as subjective to removal. The list $L_{k'}^{sim}(c)$ is pruned by removing from it all WordNet concepts with too large a s_{δ}. By applying this algorithm on the results in Table 7 (fixing k' at 4), we are able to prune out some problematic concepts, such as the WordNet concept SIDEWALK w.r.t. the LSCOM concepts COMPUTER, SPEAKER_AT_PODIUM and SPORTS, the WordNet concept ROAD with respect to the LSCOM concept COMPUTER, or the WordNet concepts PLANT and WHEEL with respect to the LSCOM concept PEDESTRIAN_ZONE. Similar results are achieved based on the results obtained by the n'-TF measure (Table 8).

Built-in bi-modality matching with GBM. Table 9 contains the results of the built-in bi-modality matching by the GBM approach in which the two modalities have been used as an integral part of the matching process. As we can see, the obtained results are in general coherent, although less performant as compared to the VSBM approach. This flaw can be attributed to the low number of concepts resulting in too little nodes in the multi-modal graph, which decreases the probability of discovering interesting matches. The performance of the matching procedure can be significantly improved by adding more concepts and increasing the number of NN-links. An overall advantage of this method is the computational time of the RWR algorithm and the multi-modality which allows the concepts to be populated by documents of different types. These two points make this method very promising for a matching at a larger scale.

5 Conclusions

The problem of associating high level meaning to a set of visual concepts has been situated in an ontology matching framework. We have proposed and compared two generic matching techniques - one based on a variable selection method (VSBM) and one based on a random walk in a graph (GBM) by relying on instances of both *visual* and *textual* nature. We have demonstrated that these two extensional modalities are complementary and their combined use improves the achieved results. Although for the moment the VSBM outperforms the GBM approach, the full potential of the latter method is to be uncovered in a large-scale application which is a subject of near future work.

The achieved alignments allow for the semantic enrichment of concepts belonging to a multimedia ontology (LSCOM) with high level linguistic concepts from a general and common sense knowledge base (WordNet). This alignment could be used to build a linguistic description of the concepts of LSCOM and improve the retrieval process through: (a) query expansion and reformulation, i.e. retrieving documents annotated with concepts from an ontology O using a query composed of concepts of an ontology O', and (b) a better description of the documents in the indexing process.

Due to the bias of the data and to the difficulty to extract the concrete semantics of a correlation, a quantitative measure of the efficiency of the approach is difficult to give. An evaluation of the approach could be envisaged within a concrete application context in an information access framework.

Table 1. Visual VSBM matching with Spearman's correlation measure (eq. (3))

Bus	Computer	Ped. Zone	Speaker At Pod.	Sport
0.602 head	0.646 person	0.726 plant	0.594 person	0.631 head
0.598 road	0.643 sidewalk	0.709 grass	0.532 head	0.613 sidewalk
0.588 car	0.636 head	0.694 wheel	0.522 grass	0.607 person
0.587 person	0.565 road	0.687 tree	0.521 plant	0.607 car
0.584 wheel	0.495 car	0.617 arm	0.503 sidewalk	0.580 road
0.581 arm	0.467 arm	0.575 leg	0.481 road	0.555 arm
0.570 tree	0.427 wheel	0.567 car	0.475 wheel	0.505 wheel
0.557 sidewalk	0.422 leg	0.478 road	0.468 tree	0.504 tree
0.552 grass	0.411 grass	0.477 sidewalk	0.383 arm	0.454 leg
0.542 plant	0.408 tree	0.467 torso	0.363 car	0.444 torso
0.509 leg	0.406 plant	0.440 person	0.341 leg	0.426 plant
0.460 torso	0.388 torso	0.413 head	0.233 torso	0.399 grass
0.336 plate	0.204 plate	0.117 plate	0.188 plate	0.320 plate

Table 2. Visual VSBM matching with n'-TF measure (eq. (4)). $n' = 150$.

Bus	Computer	Ped. Zone	Speaker At Pod.	Sport
0.325 person	0.400 head	0.600 grass	0.456 person	0.306 road
0.318 grass	0.375 person	0.531 plant	0.400 grass	0.300 head
0.318 road	0.318 sidewalk	0.475 tree	0.368 head	0.300 person
0.293 head	0.306 road	0.456 wheel	0.350 plant	0.281 sidewalk
0.268 plant	0.231 torso	0.275 leg	0.325 road	0.250 car
0.268 tree	0.225 leg	0.250 plate	0.281 sidewalk	0.212 leg
0.243 sidewalk	0.193 grass	0.212 arm	0.225 tree	0.193 arm
0.237 wheel	0.193 plant	0.162 person	0.225 wheel	0.175 torso
0.218 torso	0.118 car	0.137 car	0.162 leg	0.168 plate
0.206 leg	0.112 wheel	0.137 torso	0.106 car	0.156 plant
0.150 arm	0.100 tree	0.112 head	0.087 arm	0.143 wheel
0.150 car	0.093 arm	0.112 road	0.068 plate	0.131 tree
0.143 plate	0.087 plate	0.112 sidewalk	0.062 torso	0.118 grass

Table 3. Visual GBM matching

Bus	Computer	Ped. Zone	Speaker At Pod.	Sport
4.2E-6 car	3.4E-6 car	2.3E-6 head	2.1E-6 head	3.0E-6 head
2.9E-6 head	3.3E-6 head	2.0E-6 car	1.7E-6 car	2.5E-6 car
2.2E-6 tree	2.1E-6 tree	1.5E-6 tree	1.2E-6 tree	2.0E-6 person
1.6E-6 road	1.4E-6 person	1.0E-6 road	1.0E-6 person	1.9E-6 tree
1.4E-6 person	1.4E-6 road	8.6E-7 person	7.9E-7 road	1.0E-6 road
9.6E-7 sidewalk	8.9E-7 sidewalk	7.3E-7 sidewalk	4.7E-7 sidewalk	9.9E-7 plant
7.3E-7 plant	7.8E-7 wheel	5.6E-7 arm	4.6E-7 grass	7.6E-7 sidewalk
6.2E-7 arm	6.6E-7 plant	5.5E-7 leg	3.3E-7 wheel	6.9E-7 grass
5.4E-7 grass	6.4E-7 grass	5.5E-7 torso	3.3E-7 plant	5.7E-7 wheel
5.3E-7 leg	5.8E-7 plate	4.8E-7 grass	2.9E-7 arm	4.7E-7 plate
5.1E-7 wheel	2.6E-7 arm	4.6E-7 wheel	2.4E-7 torso	3.4E-7 arm
5.1E-7 torso	2.4E-7 leg	2.4E-7 plant	2.4E-7 leg	3.4E-7 leg
3.1E-7 plate	2.4E-7 torso	1.6E-7 plate	2.3E-7 plate	3.3E-7 torso

Table 4. Textual VSBM matching with Spearman's correlation measure (eq. (3))

Bus	Computer	Ped. Zone	Speaker At Pod.	Sport
0.667 head	0.531 head	0.615 head	0.485 head	0.581 head
0.363 car	0.166 car	0.336 car	0.109 car	0.223 car
0.233 tree	0.059 person	0.234 tree	0.006 tree	0.170 tree
0.153 person	0.050 tree	0.157 person	0.011 person	0.118 person
0.130 road	0.013 torso	0.125 road	0.053 torso	0.070 torso
0.098 arm	0.005 arm	0.109 arm	0.068 arm	0.050 arm
0.092 torso	0.023 leg	0.104 grass	0.113 leg	0.042 grass
0.086 grass	0.094 plate	0.096 torso	0.170 road	0.023 road
0.069 leg	0.101 plant	0.068 leg	0.185 grass	0.017 leg
0.015 sidewalk	0.124 grass	0.017 sidewalk	0.186 plate	0.067 sidewalk
0.044 plant	0.137 road	0.066 plant	0.198 plant	0.108 plant
0.059 wheel	0.184 sidewalk	0.070 wheel	0.249 sidewalk	0.122 plate
0.091 plate	0.305 wheel	0.136 plate	0.380 wheel	0.194 wheel

Table 5. Textual VSBM matching with the n'-TF measure (eq. (4)). $n' = 150$.

Bus	Computer	Ped. Zone	Speaker At Pod.	Sport
0.257 road	0.257 person	0.185 sidewalk	0.200 person	0.228 grass
0.185 car	0.185 arm	0.185 road	0.157 plant	0.200 tree
0.171 wheel	0.171 torso	0.171 person	0.157 torso	0.171 road
0.171 sidewalk	0.171 leg	0.171 car	0.142 leg	0.157 plant
0.157 tree	0.157 tree	0.171 tree	0.100 arm	0.142 person
0.142 grass	0.142 plate	0.171 grass	0.100 grass	0.128 sidewalk
0.128 person	0.114 car	0.157 wheel	0.057 head	0.114 arm
0.114 plant	0.114 head	0.114 arm	0.057 plate	0.114 plate
0.100 head	0.100 road	0.114 head	0.057 sidewalk	0.114 torso
0.085 arm	0.100 plant	0.114 leg	0.057 tree	0.100 car
0.085 plate	0.085 grass	0.100 plant	0.042 car	0.100 head
0.085 torso	0.071 wheel	0.100 torso	0.042 road	0.100 leg
0.071 leg	0.028 sidewalk	0.028 plate	0.014 wheel	0.057 wheel

Table 6. Textual GBM matching

Bus	Computer	Ped. Zone	Speaker At Pod.	Sport
2.4E-6 road	1.0E-6 person	2.5E-6 road	5.6E-7 head	7.0E-7 grass
1.8E-6 tree	6.9E-7 head	2.2E-6 sidewalk	4.2E-7 person	6.0E-7 tree
1.5E-6 person	6.8E-7 arm	1.3E-6 tree	3.7E-7 tree	5.9E-7 person
1.4E-6 wheel	6.5E-7 leg	1.2E-6 car	3.4E-7 arm	5.6E-7 road
1.4E-6 plant	6.4E-7 torso	7.6E-7 wheel	3.3E-7 torso	4.3E-7 head
1.3E-6 car	5.4E-7 plant	6.2E-7 person	3.0E-7 leg	4.1E-7 torso
8.9E-7 sidewalk	2.5E-7 tree	4.4E-7 arm	2.8E-7 plant	4.1E-7 leg
8.3E-7 leg	2.1E-7 wheel	4.2E-7 head	1.6E-7 plate	4.0E-7 plant
7.3E-7 arm	1.9E-7 road	3.6E-7 leg	9.5E-8 grass	4.0E-7 arm
7.2E-7 head	1.6E-7 plate	3.5E-7 torso	9.4E-8 road	2.3E-7 sidewalk
6.7E-7 torso	1.3E-7 sidewalk	3.4E-7 plant	8.1E-8 wheel	1.8E-7 car
5.1E-7 grass	9.6E-8 grass	3.3E-7 grass	6.0E-8 sidewalk	1.7E-7 wheel
4.8E-7 plate	9.1E-8 car	1.2E-7 plate	5.9E-8 car	7.7E-8 plate

Table 7. Differences between the visual and the textual similarity scores issued from the VSBM matching with Spearman's correlation measure

Bus	s_v s_t	s_δ	Comp.	s_v s_t	s_δ	Ped.Zone	s_v s_t	s_δ	Sp.AtPod.	s_v s_t	s_δ	Sport	s_v s_t	s_δ
head	1 1	0	person	1 3	2	plant	1 11	10	person	1 4	3	head	1 1	0
road	2 5	3	sidewalk	2 12	10	grass	2 7	5	head	2 1	1	sidewalk	2 10	8
car	3 2	1	head	3 1	2	wheel	3 12	9	grass	3 9	6	person	3 4	1
person	4 4	0	road	4 11	7	tree	4 3	1	plant	4 11	7	car	4 2	2
wheel	5 12	7	car	5 2	3	arm	5 6	1	sidewalk	5 12	7	road	5 8	3
arm	6 6	0	arm	6 6	0	leg	6 9	3	road	6 8	2	arm	6 6	0
tree	7 3	4	wheel	7 13	6	car	7 2	5	wheel	7 13	6	wheel	7 13	6
sidewalk	8 10	2	leg	8 7	1	road	8 5	3	tree	8 3	5	tree	8 3	5
grass	9 8	1	grass	9 10	1	sidewalk	9 10	1	arm	9 6	3	leg	9 9	0
plant	10 11	1	tree	10 4	6	torso	10 8	2	car	10 2	8	torso	10 5	5
leg	11 9	2	plant	11 9	2	person	11 4	7	leg	11 7	4	plant	11 11	0
torso	12 7	5	torso	12 5	7	head	12 1	11	torso	12 5	7	grass	12 7	5
plate	13 13	0	plate	13 8	5	plate	13 13	0	plate	13 10	3	plate	13 12	1

Table 8. Differences between the visual and the textual similarity scores issued from the VSBM matching with the n'-TF similarity measure. $n' = 150$.

Bus	s_v s_t	s_δ	Comp.	s_v s_t	s_δ	Ped.Zone	s_v s_t	s_δ	Sp.AtPod.	s_v s_t	s_δ	Sport	s_v s_t	s_δ
person	1 6	4	head	1 6	4	grass	1 2	1	person	1 1	0	road	1 3	2
road	2 1	1	person	2 1	1	plant	2 5	3	grass	2 4	2	head	2 8	6
grass	2 5	3	sidewalk	3 10	7	tree	3 2	1	head	3 5	2	person	3 5	2
sidewalk	3 3	0	road	4 7	3	wheel	4 3	1	plant	4 2	2	sidewalk	4 6	2
tree	3 4	1	torso	5 3	2	leg	5 4	1	road	5 6	1	car	5 8	3
plant	3 7	4	leg	6 3	3	plate	6 6	0	sidewalk	6 5	1	leg	5 9	4
head	3 8	5	grass	7 8	1	arm	7 4	3	tree	7 5	2	arm	6 7	1
wheel	4 3	1	plant	7 7	0	person	8 2	6	wheel	8 7	1	torso	7 7	0
torso	5 9	4	car	8 6	2	car	9 2	7	leg	9 3	6	plate	8 7	1
leg	6 10	4	wheel	8 9	1	torso	9 5	4	car	10 6	4	plant	9 4	5
car	7 2	5	tree	9 4	5	head	10 4	6	arm	11 4	7	wheel	10 10	0
arm	7 9	2	arm	10 2	8	road	10 1	9	plate	12 5	8	tree	11 2	9
plate	8 9	1	plate	11 5	6	sidewalk	10 1	9	torso	13 2	11	grass	12 1	11

Table 9. A built-in bi-modality matching with GBM

Bus	Computer	Ped. Zone	Speaker At Pod.	Sport
3.8E-6 car	2.6E-6 head	2.7E-6 road	1.8E-6 head	2.2E-6 head
3.0E-6 road	2.2E-6 car	2.4E-6 sidewalk	1.2E-6 car	1.8E-6 tree
2.9E-6 tree	1.6E-6 tree	2.4E-6 car	1.1E-6 tree	1.8E-6 car
2.4E-6 head	1.6E-6 person	2.1E-6 tree	1.0E-6 person	1.7E-6 person
2.0E-6 person	1.1E-6 road	1.9E-6 head	6.7E-7 road	1.1E-6 road
1.5E-6 plant	8.9E-7 plant	1.0E-6 person	5.0E-7 arm	1.0E-6 grass
1.4E-6 wheel	7.7E-7 sidewalk	1.0E-6 wheel	4.7E-7 plant	1.0E-6 plant
1.4E-6 sidewalk	7.4E-7 arm	8.0E-7 arm	4.6E-7 torso	7.6E-7 sidewalk
1.0E-6 arm	7.0E-7 leg	7.4E-7 leg	4.4E-7 leg	5.9E-7 leg
1.0E-6 leg	6.9E-7 torso	7.4E-7 torso	4.2E-7 grass	5.9E-7 arm
9.4E-7 torso	6.8E-7 wheel	6.4E-7 grass	4.1E-7 sidewalk	5.8E-7 torso
7.9E-7 grass	5.4E-7 plate	4.7E-7 plant	3.1E-7 plate	5.5E-7 wheel
5.9E-7 plate	5.3E-7 grass	2.2E-7 plate	3.1E-7 wheel	4.1E-7 plate

References

1. Athanasiadis, T., Tzouvaras, V., Petridis, K., Precioso, F., Avrithis, Y., Kompatsiaris, Y.: Using a multimedia ontology infrastructure for semantic annotation of multimedia content. In: SemAnnot 2005 (2005)
2. Dasiopoulou, S., Kompatsiaris, I., Strintzis, M.: Using fuzzy dls to enhance semantic image analysis. In: Semantic Multimedia, pp. 31–46. Springer, Heidelberg (2008)
3. Dasiopoulou, S., Tzouvaras, V., Kompatsiaris, I., Strintzis, M.: Enquiring MPEG-7 based multimedia ontologies. In: MM Tools and Appls., pp. 1–40 (2010)
4. Deng, J., Dong, W., Socher, R., Li, L., Li, K., Fei-Fei, L.: ImageNet: a large-scale hierarchical image database. In: CVPR, pp. 710–719 (2009)
5. Euzenat, J., Shvaiko, P.: Ontology Matching, 1st edn. Springer, Heidelberg (2007)
6. Fan, J., Luo, H., Shen, Y., Yang, C.: Integrating visual and semantic contexts for topic network generation and word sense disambiguation. In: ACM CIVR 2009, pp. 1–8 (2009)
7. Guyon, I., Elisseeff, A.: An introduction to variable and feature selection. JMLR 3(1), 1157–1182 (2003)
8. Haveliwala, T.: Topic-sensitive pagerank: A context-sensitive ranking algorithm for web search. IEEE Transactions on Knowledge and Data Engineering, 784–796 (2003)
9. Hudelot, C., Atif, J., Bloch, I.: Fuzzy Spatial Relation Ontology for Image Interpretation. Fuzzy Sets and Systems 159, 1929–1951 (2008)
10. Hudelot, C., Maillot, N., Thonnat, M.: Symbol grounding for semantic image interpretation: from image data to semantics. In: SKCV-Workshop, ICCV (2005)
11. Inoue, M.: On the need for annotation-based image retrieval. In: Proceedings of the Workshop on Information Retrieval in Context (IRiX), Sheffield, UK, pp. 44–46 (2004)
12. James, N., Todorov, K., Hudelot, C.: Ontology matching for the semantic annotation of images. In: FUZZ-IEEE. IEEE Computer Society Press, Los Alamitos (2010)

13. Koskela, M., Smeaton, A.: An empirical study of inter-concept similarities in multimedia ontologies. In: CIVR 2007, pp. 464–471. ACM, New York (2007)
14. Mihalcea, R., Tarau, P., Figa, E.: Pagerank on semantic networks, with application to word sense disambiguation. In: ICCL, p. 1126. Association for Computational Linguistics (2004)
15. Miller, G.: WordNet: a lexical database for English. Communications of the ACM 38(11), 39–41 (1995)
16. Pan, J., Yang, H., Faloutsos, C., Duygulu, P.: Automatic multimedia cross-modal correlation discovery. In: ACM SIGKDD, p. 658. ACM, New York (2004)
17. Peraldi, I.S.E., Kaya, A., Möller, R.: Formalizing multimedia interpretation based on abduction over description logic aboxes. In: Description Logics (2009)
18. Russell, B., Torralba, A., Murphy, K., Freeman, W.: LabelMe: a database and web-based tool for image annotation. IJCV 77(1), 157–173 (2008)
19. Smeulders, A., Worring, M., Santini, S., Gupta, A., Jain, R.: Content-based image retrieval at the end of the early years. IEEE Trans. Patt. An. Mach. Intell., 1349–1380 (2000)
20. Smith, J., Chang, S.: Large-scale concept ontology for multimedia. IEEE Multimedia 13(3), 86–91 (2006)
21. Snoek, C., Huurnink, B., Hollink, L., De Rijke, M., Schreiber, G., Worring, M.: Adding semantics to detectors for video retrieval. IEEE Trans. on Mult. 9(5), 975–986 (2007)
22. Tansley, R.: The multimedia thesaurus: An aid for multimedia information retrieval and navigation. Master's thesis (1998)
23. Todorov, K., Geibel, P., Kühnberger, K.-U.: Extensional ontology matching with variable selection for support vector machines. In: CISIS, pp. 962–968. IEEE Computer Society Press, Los Alamitos (2010)
24. Tong, H., Faloutsos, C., Pan, J.-Y.: Fast random walk with restart and its applications. In: ICDM 2006, pp. 613–622. IEEE Computer Society, Washington, DC (2006)
25. Wang, C., Jing, F., Zhang, L., Zhang, H.: Image annotation refinement using random walk with restarts. In: ACM MM, p. 650 (2006)
26. Wu, L., Hua, X.-S., Yu, N., Ma, W.-Y., Li, S.: Flickr distance. In: MM 2008, pp. 31–40. ACM, New York (2008)
27. Yang, Y., Pedersen, J.O.: A comparative study on feature selection in text categorization. In: Fourteenth ICML, pp. 412–420. Morgan Kaufmann Publishers, San Francisco (1997)
28. Yao, B., Yang, X., Lin, L., Lee, M., Zhu, S.: I2t: Image parsing to text description. IEEE Proc. Special Issue on Internet Vision (to appear)

Automated Annotation of Landmark Images Using Community Contributed Datasets and Web Resources

Gareth J.F. Jones[1], Daragh Byrne[1,2], Mark Hughes[1,2], Noel E. O'Connor[2], and Andrew Salway[1]

[1] Centre for Digital Video Processing, School of Computing, Dublin City University, Dublin 9, Ireland
[2] CLARITY: Centre for Sensor Web Technologies, Dublin City University, Dublin 9, Ireland
{gjones,dbyrne,mhughes,asalway}@computing.dcu.ie, Noel.OConnor@dcu.ie

Abstract. A novel solution to the challenge of automatic image annotation is described. Given an image with GPS data of its location of capture, our system returns a semantically-rich annotation comprising tags which both identify the landmark in the image, and provide an interesting fact about it, e.g. "A view of the Eiffel Tower, which was built in 1889 for an international exhibition in Paris". This exploits visual and textual web mining in combination with content-based image analysis and natural language processing. In the first stage, an input image is matched to a set of community contributed images (with keyword tags) on the basis of its GPS information and image classification techniques. The depicted landmark is inferred from the keyword tags for the matched set. The system then takes advantage of the information written about landmarks available on the web at large to extract a fact about the landmark in the image. We report component evaluation results from an implementation of our solution on a mobile device. Image localisation and matching offers 93.6% classification accuracy; the selection of appropriate tags for use in annotation performs well (F1M of 0.59), and it subsequently automatically identifies a correct toponym for use in captioning and fact extraction in 69.0% of the tested cases; finally the fact extraction returns an interesting caption in 78% of cases.

Keywords: web mining, geo-tagged images, landmark identification, automated image captioning.

1 Introduction

Photo capture, storage and usage has undergone a revolution in recent years. Many people routinely take large numbers of images from their lives using dedicated digital cameras, and increasingly with those embedded in mobile devices such as smartphones. Many of the devices used for image capture now incorporate GPS sensors meaning that the location at which an image was captured is

T. Declerck et al. (Eds.): SAMT 2010, LNCS 6725, pp. 111–126, 2011.

easily available. These images are then very often shared with others using online photo archives such as Flickr or Facebook. Users uploading images are expected to provide captions for their images which must be entered manually. However, there are disadvantages with this manual annotation process. First, those taking pictures will often be traveling in places which they do not know well and so are not able to provide accurate and/or interesting labels. Thus images taken while on a visit to London may simply be labeled "London". The volume of images taken on such a trip means that even if they are knowledgeable about the place being visited, users will often not take the time to provide detailed captions, and even if they do this, labels will be inconsistent between different users uploading to social repositories, reducing the effectiveness of subsequent image search. In this paper we describe a mobile application running on an iPhone in conjunction with a web service which automates this captioning process for landmark images. Once captioned images can be uploaded to an online social media application. We also believe our work is relevant in the context of the burgeoning interest in augmented reality whereby a camera screen on a mobile device is automatically supplemented with caption-like details about the target image.

Our approach exploits GPS information accompanying an image, geographic resources that provide reverse lookup, e.g. GeoNames [1], the existing keyword tags associated with images in community contributed datasets such as Flickr, and the information about a great many places available taken from the World Wide Web. Given a GPS-tagged image of a landmark our system can generate a caption comprising keyword tags that describe what landmark is in the image and give an interesting fact about it. Classification is achieved through the integration of image classification based on computational classification methods (using local image features [4] and Support Vector Machines [6]) and a technique for text information extraction from the web.

This paper is organised as follows: Section 2 reviews background to our work, Section 3 describes the component stages of our system and how they are integrated to generate image captions, Section 4 describes the application and evaluation of the components, and finally Section 5 concludes our work to date.

2 Background

The 'semantic gap' between the low-level features used in many content-based image analysis technologies and the human interpretation of images means that most large-scale image retrieval systems in use today are based on textual metadata in the form of tags or captions which are usually created manually by humans. Thus the potential for retrieved images to satisfy the user relies heavily on the accuracy and quality of these human-defined tags. The main disadvantage of this approach is that these manual tags will be inconsistent and often poor due to lack of knowledge or time on the part of the annotator.

An alternative and more appealing image search scenario in many situations is to use a query image which should be compared to existing known images in order to identify what is depicted in the new image. High-level semantic classification can be used to identify complex images such as the image being a view of the Trevi Fountain in Rome. A popular approach to automatic high-level semantic classification is to use local image features as opposed to low-level image features. Local image features are based around interest points (salient, non uniform regions) in a photo. These have been successfully used in the past for object matching, tracking and recognition along with other niche tasks such as image mosaicing [9]. Several research groups have used these local image features for object recognition tasks on mobile devices. For example Chevallet et al.[5] developed an application called Snap2Tell designed to run on GPS enabled mobile phone devices. In their system a user can take a picture of one of 120 landmarks located around Singapore (STOIC dataset), which is then identified on a remote server. Fritz et al. [7] use SIFT features for descriptor matching in a mobile landmark recognition system. They use a relatively small dataset of 1005 landmark images (ZUBUD dataset) based around Zurich as their training collection. When a user takes a picture of a landmark within the Zurich region, the system aims to classify the image against the dataset using only the SIFT features that it deems to be 'informative', disregarding all other features to speed up matching time. Yeh et al. [19] developed a system that recognises an image of a landmark taken on a mobile device and retrieves information from an online search describing the landmark within an image. This system compares an image of a landmark taken on a mobile device against a collection of images that are contained on webpages. If a match is found, the system extracts information from the text in the corresponding webpage and uses it to extract information from the wider web to describe the landmark depicted within the image.

Our work improves on this previous work by implementing a more robust recognition system that is able to classify landmarks more accurately using very large datasets of community contributed images. Our techniques allow for the creation of reliable captions and tags from large amounts of noisy data. The integrated framework then reliably augments these captions and tags with facts about landmarks contained within an image taken from the whole Web.

3 Landmark Identification and Caption Generation

Our landmark identification and captioning application exploits the very large number of captioned geo-tagged images now available in community contributed image collections from which tags are selected. In this integrated process when a new geo-tagged image is introduced into the system, a cluster of similar images is first identified. Representative sets of tags for this landmark based on a set of matched images are then identified and used to identify the primary toponym in the image. Finally keywords selected from images related to the one being captioned are used to extract a fact about the landmark from the web and integrate it into the image's caption.

3.1 Landmark Classification

An effective approach to classification of a landmark image is to harvest a large number of similarly annotated landmark images, and then to match it based on context and content features of these images [14] [15]. In this process image and object matching using interest point features has been shown to work well even in large-scale image databases containing thousands of different images [11]. However the actual matching between keypoints can be very computationally expensive, and for large-scale image databases containing millions of images computationally infeasible. In order to use these techniques in a practical system, methods are required to reduce the number of keypoints which need to be compared or else that do not match keypoint to keypoint. In our work this is achieved by combining computer vision techniques with different forms of semantic context data to organise and classify landmarks within images.

A new framework is implemented based on *single viewpoint clustering* [10] which enables the efficient and accurate classification of landmarks using a large scale training database. Single viewpoint clustering involves collecting a number of images of the same landmark taken from a relatively similar viewpoint and clustering them to create clusters of visually similar images. Each cluster can then be assigned spatial location data. They can be used for efficient classification of new input images using Support Vector Machines (SVMs), where each SVM model represents a single landmark from a certain viewpoint. One drawback of this approach is that a large number of positive examples are needed to train an accurate SVM classifier. Within real-world collections of landmark images, there may not be a sufficient number of images to build a reliable SVM model. To deal with this situation, we use a method that addresses this problem by combining SVMs with an hierarchical classification approach.

In this paper we apply this technique to a collection of images harvested from Flickr which contains many examples of images for significant landmarks. For example a search on Flickr for "eiffel tower" currently returns over 370,000 images and for "notre dame paris" over 245,000 images (May 2010). Landmarks tend to have a unique visual appearance that leads to high discrimination values between different landmarks. The automated classification approach applied here works well due to the observed capture behaviour of users on large scale photo-sharing websites. Photographers tend to visit similar destinations and landmarks, and to take images of these landmarks from a small number of locations due to geographical constraints and their photogencity from certain viewpoints. This leads to a large overlap of visually similar images of popular landmarks. Based on this observation, our system takes advantage of this overlap by reducing the search space in a large scale dataset by clustering similar images, thus creating a robust means of classifying an image using SVMs.

Dataset. For this work a training collection of images was harvested using the Flickr API from the metropolitan area of Paris. In order to reliably cluster similar images for SVM training, we downloaded only geo-tagged images. To ensure that the vast majority of these images contained large landmarks as their main subject, the image text tags assigned to describe it when it was uploaded to Flickr were analysed. A long list of stopwords was created to filter out unwanted images (eg. concert, match, march). This filtering process produced a training set of 76,749 geo-tagged images. However, since geo-tags and text tags are assigned by those contributing the images to Flickr there is no way to ensure their accuracy. It is however expected that correct tags will dominate the dataset.

SVM Classification. Single viewpoint clustering involves taking a number of images of the same landmark taken from a relatively similar viewpoint, and clustering them into visually similar clusters. Due to the large size of the dataset, traditional clustering techniques such as K-means would be infeasible. We explored many different techniques and combinations of image features (content and context) to determine an efficient and accurate method to cluster large numbers of images. We selected a combination of spatial data, low-level image features and SURF local image features [4].

All training images were first clustered based on geographical locations. Cluster centres were chosen randomly in the dataset and all images located within a spatial radius of 500 metres of each centre were clustered. This process was repeated until all images in the dataset were assigned to a cluster. Within each of these clusters, images were then subclustered based on two MPEG7 low-level features: edge histogram (which provides weak spatial verification) and scalable colour using an hierarchical clustering method.

Each of these clusters were then subclustered again based on local image feature matching. A graph was created using images within a cluster as the nodes and local feature matched as the edges. The most connected image was chosen as the cluster centre. All images were then compared against the cluster centres to subcluster these images into visually similar clusters. Thus, each cluster should represent a landmark from a similar viewpoint. Clusters were then assigned a spatial location based on the average position of each image within the cluster.

K-means clustering was then carried out on these clusters (K = 100) using their geographical location as the comparison values. The metropolitan area of Paris was split into 100 geographical regions and a multi-class SVM model trained to represent all classifiable landmarks within each of these regions. Thus 100 multi-class classification models were trained in total, each one representing all viewpoints of landmarks within the geographical bounding box as determined by the K-means clustering procedure. These models were trained using Visual Bag of Words (BOW) features with a vocabulary size of 4096. We also trained versions of these SVM models using the MPEG7 edge histogram descriptor.

To classify an image, firstly its closest multi-class SVM is retrieved based on geographical distance. A visual BOW is created for the test image based on the vocabulary used for the training images. The SVM then classifies the test input vector into one of the classes used to create the model. At this point an input image is only classified to the nearest class so a more definite verification is required to guarantee an accurate match. The input image is then compared against all images within this class using point to point matching with SURF image features. If the number of matches is above a threshold, this should confirm that the input image is indeed a match.

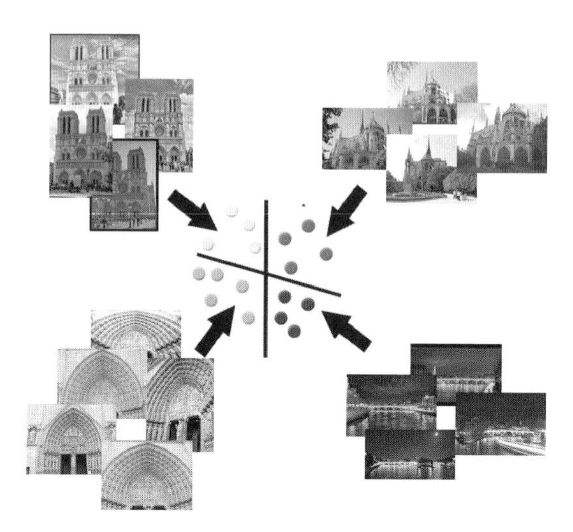

Fig. 1. Example of the SVM training process within a small spatial area in Paris. All clusters within this geographical area are used as inputs into a multi-class SVM model.

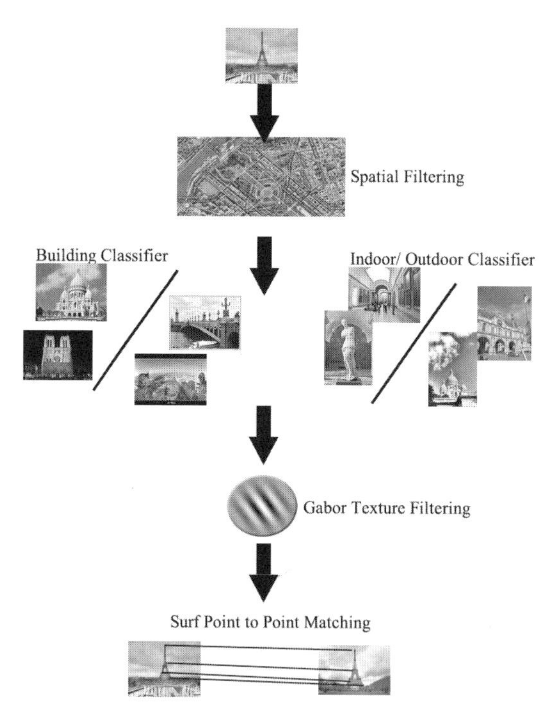

Fig. 2. The hierarchical classification pipeline. Images that cannot be successfully classified using the spatially organised SVM models are classified using this approach.

A significant drawback of this approach is that a large number of training examples are needed to build each accurate SVM model. Our hybrid approach addresses this problem as follows, in cases where no match is found for an input image using the SVM approach, a slower hierarchical pipeline classification method is used.

Hierarchical Classification. To classify an input image using the hierarchical approach all images within a spatial radius are first retrieved. Many different spatial radii were investigated. It was found that for our urban Paris dataset that a radius of 500 meters provided the best trade-off between accuracy and speed. Although high level semantics based on image features are very difficult to implement successfully, several low-level semantic classifiers can work quite well. In our work two low-level semantic classifiers that have been shown to work well in the past are indoor/outdoor and building/non-building [17][12], were used in the early phase of our hierarchical pipeline. Classifiers were trained based on MPEG7 features using SVMs to classify whether an image was taken indoors or outdoors and whether an image contains a large building. The number of retrieved images is then pruned based on the results of these semantic classifiers. Gabor texture features are then extracted and compared against the remainder of the retrieved images. All images which have a Euclidean distance above a threshold (threshold = 20) are then pruned from the search space. Point to point matching is then carried out using SURF image features and the distance ratio test. The image with the highest number of positive matches (if above a threshold of 4) is then matched with the input image

3.2 Landmark Identification and Selection of Keywords, Tags and Image Title

The landmark image localization and matching method described in the previous section was implemented into a matching engine and service. The engine localizes and matches the input image, looking up potentially relevant images within the training dataset of Parisian landmark images. Once potential matches have been located, the engine returns a set of images from the training set ordered from best match to worst in XML format. For each image returned, associated metadata is also provided, including the community-contributed tags, the title given to the image by the uploader and a unique identifier for the matched image.

Tag Filtering And Selection. Each matched image returned by the localization engine has associated information which includes a set of tags. We wish to assign the most appropriate tags to the input image from among those contributed by the community. Thus our system attempts to identify a representative set of tags which can be used to annotate the target image. A hashmap of tags is created from the available tag set, by iterating through the matched images and the tags they contain and adding them to the set. A tag within the set is given an importance measure or weighting which is incremented with each encounter. The amount by which it is incremented corresponds to the rank of the matched image within the results list, i.e. tags encountered in the highest scoring match are given greater weight than those from images further down the list. Finally, using the weighted score of each tag, the set is thresholded to yield the set of tags likely to be the most representative of the target image.

Toponym Identification. Using the provided location information, our application middleware communicates with the GeoNames API and retrieves a list of toponyms within a 1.5 kilometre radius of these coordinates. A wide radius is employed to allow for positioning errors or varying ranges of accuracy in the provided GPS position. The toponym list is filtered to remove irrelevant, spurious or noisy toponym types, e.g. names of hotels. The list of potential toponyms is then matched as outlined above to the selected tags to identify the most likely candidate. This is then applied as the image title. With a toponym identified, the fact extraction and caption augmentation service, as described below, is called and the returned facts are then used to further annotate the media being processed. We employ a reasonably straightforward approach: the available toponyms are compared to the thresholded tag set and the best matching item chosen. Stopwords are removed from the titles, which are divided into tokens and stemmed using the Porter Stemming algorithm [13]. Using the Jaccard Coefficient [18], each toponym is then compared to the selected tags, and scored. The best match is then returned, or if no match is found, the closest toponym is returned.

3.3 Fact Extraction and Title Augmentation

In the next stage of processing we use the output of the image classification and toponym identification as input to a highly portable mechanism for the extraction of partially structured facts from information on toponyms available on the World Wide Web. A particular feature of this is that it exploits information redundancy on the web, i.e. the fact that the same information about a landmark is available in many forms on the web. This method is described in detail in [16]. For a given landmark, we return a list of facts in the form (Landmark, Cue, Text-Fragment), ranked according to a score which is intended to promote interesting and true facts. This fact structure makes it straightforward to combine it with an existing image title. Crucially, for this information extraction process, we assume that at least one key fact about a landmark will be expressed somewhere on the web in a simple form, so that we only need to work with a few simple linguistic structures and shallow language processing. The following sub-sections describe the fact extraction process.

Get Snippets from Search Engine: A series of queries is made to a web search engine (we use Yahoo's BOSS API [3]). Each query takes the form <"Landmark Cue">; where the use of double quotes indicates that only exact matches are wanted, i.e. text in which the given landmark and cue are adjacent. A set of cues is manually specified to capture some common and simple ways in which information about landmarks is expressed, e.g. 'is a', 'is famous for', 'is popular with', 'was built'.

Although we worked with around 40 cues (including single / plural and present / past forms), a much smaller number are responsible for returning the majority of high ranking facts; in particular (and perhaps unsurprisingly) the generic "is" seems most productive. The query may also include a disambiguating term. For example, streets and buildings with the same name may occur in different towns, so we can include a town name in the query outside the double quotes, e.g. <"West Street is popular with" Bridport>. For each query, all the unique snippets returned up to a preconfigured maximum number are processed in the next step. Typically a snippet is a few lines of text from a webpage around the words that match the query, often broken in mid-sentence.

Shallow Chunk Snippets to Make Candidate Facts: Because we are only retrieving information about a given landmark that is expressed as "Landmark Cue ...", we can use a simple extraction pattern to obtain candidate facts from the retrieved snippets. The gist of the pattern is 'BOUNDARY LANDMARK CUE TEXT-FRAGMENT BOUNDARY', such that 'TEXT-FRAGMENT' captures the 'Text-Fragment' part of a fact. The details of the pattern are captured in a regular expression on a language-specific basis, e.g. to specify boundary words and punctuation, to allow optional words to appear inbetween LANDMARK and CUE, and to reorder the elements for non-SVO languages. A successful match of the pattern on a snippet leads to the generation of a candidate fact. For example, using extraction patterns the snippet text '...in London. Big Ben was named after Sir Benjamin Hall. ...' matches, giving the candidate fact (Big Ben, was named, after Sir Benjamin Hall) but 'The square next to Big Ben was named in 1848...' does not match.

Filter Candidate Facts: Four filters are used as a quality control to remove candidate facts that: contain potentially subjective words; end in words that would be ungrammatical; are under a length threshold; and that contain words that are all in capitals. Finally, facts are ranked so that we are more likely to get correct and interesting facts at the top. We exploit the overlap between candidate facts for the same Landmark-Cue pair to capture these notions to some extent. For each Landmark-Cue pair a keyword frequency list is generated by counting the occurrence of all words in the Text-Fragments for that pair, words in a stopword list are ignored. The score for each fact is then calculated by summing the Landmark-Cue frequencies of each word in the Text-Fragment, so that facts containing words that were common in other facts with the same Landmark-Cue will score highly. If shorter facts are wanted then the sum is divided by the word length of the Text-Fragment.

The sum score for a fact can become high in two ways: (i) there are many overlapping Text-Fragments for an Landmark-Cue pair, so there are some high word frequencies; and (ii) a fact contains more of these high frequency words than other facts. Thus, the method is designed to highly rank facts with the most appropriate Cue for the Landmark, and the best Text-Fragment for the Landmark-Cue pair. For an existing image title, e.g. "A view of the Eiffel Tower", then the top-ranked fact, e.g. 'Eiffel Tower, was built, in 1889 for an international exhibition in Paris', can be inserted in one of two ways: (i) as a new sentence - "A view of the Eiffel Tower. The Eiffel Tower was built in 1889..."; or (ii) as a subclause - "A view of the Eiffel Tower, which was built in 1889...".

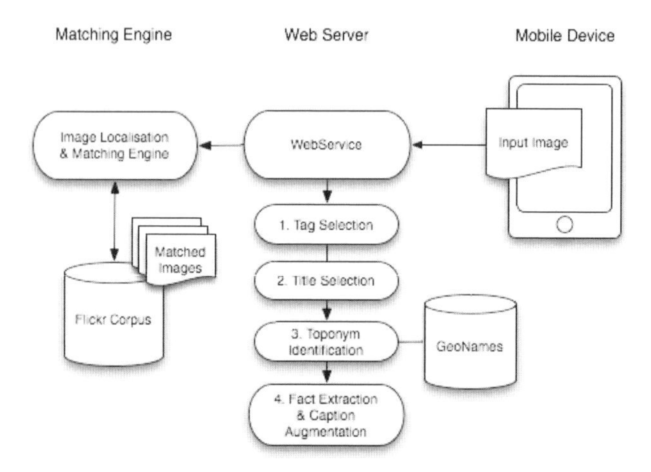

Fig. 3. An illustration demonstrating how the different components of the framework integrate with one another

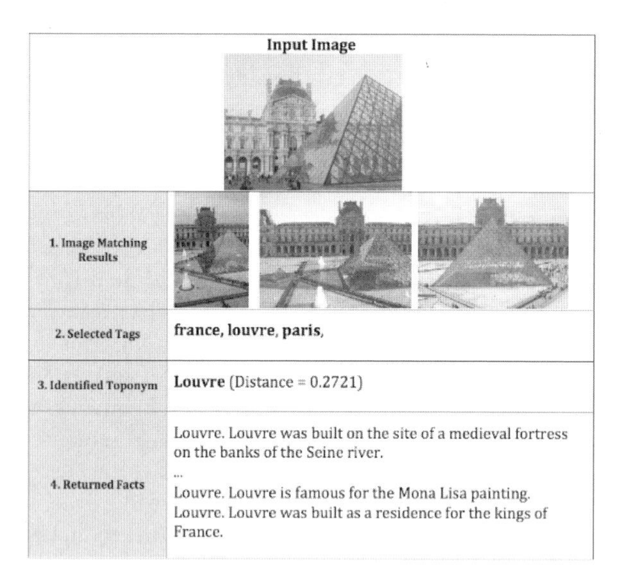

Fig. 4. The workflow and outputs of the chained components illustrated with a worked example, in this case of the Louvre in Paris

4 Application and Evaluation

4.1 Application Workflow

The three components described in the previous sections are integrated into a combined service architecture shown in Figure 3. The landmark image recognition and classification engine resides on the server along with a series of web services designed to expose their functionality to a mobile application running on a compatible mobile device (in this case an iPhone.) The integrated service allows an input landmark image to be recognized, localized and matched with other images in the repository, in this case the Flickr corpus of Parisian landmarks described previously. In order to caption and tag an input image, matches for the provided image are looked up, and through the middleware layer used to determine appropriate annotations to be applied to the target image. The steps are as follows: first the image matching is performed after which, and using the returned results along with their associated metadata, a set of representative tags for the image being processed is identified. Using these tags, the best matching toponym nearby the provided coordinates is determined, this is then used to seed the fact extraction and caption augmentation step. This workflow is illustrated using a worked example in Figure 4.

4.2 Mobile Application

The mobile application is designed to operate in-situ with a tourist style scenario in mind. Examplescreenshots of the working iPhone application are shown in Figure 5. The application operates as follows: first the user selects a photo they want to process, either by taking a new image with the device's in-built camera or by selecting an existing image from the photo library. They are then asked to confirm that the location for the image is correct, after which the image and location data is passed to the middleware layer through a REST-based API. After the service completes the matching and annotation of the image, a response is returned to the device. The annotated image is then saved to a local data store and the application presents the results on-screen. The image, along with the automatically generated captions and tags, can then be uploaded to a number of social media sites including Flickr and Twitter through the results screen.

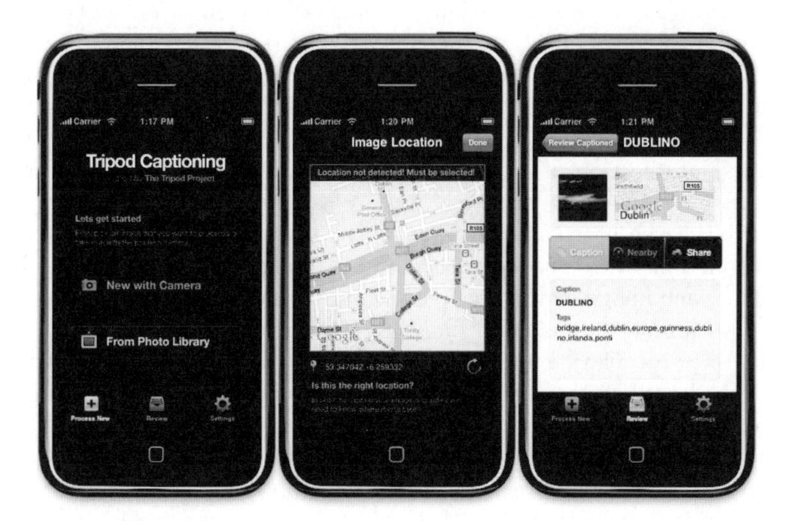

Fig. 5. An example of the application developed based on this framework running on an Apple iPhone™mobile device

Table 1. Landmark Classification Accuracy(270 test images)

Approach	No. of images	Classified correctly
Hierarchical only	270	91.0%
SVM (BOW) only	156	92.9%
SVM (Edge) only	214	93.4%
Hybrid (BOW)	270	93.3%
Hybrid (Edge)	270	93.6%

4.3 Evaluation

This section describes laboratory evaluation of the accuracy of the components of our landmark image classification and annotation system.

Landmark Classification. In order to evaluate the classification system, a test collection of 270 images was gathered from the Panoramio online collection using Panoramio's REST API [2]. All of these test images have a large landmark as their main subject and include geo-tags. It should be noted that to make the test collection realistically challenging, the landmarks shown in many of the images are partially occluded and taken under a wide variety of lighting conditions (day, night, flash, etc..). The test collection was first classified solely using the hierarchical classification approach described in section 3.1, with the aim of ascertaining the classification accuracy of this technique. Experimental results are shown in Table 1.

We then processed our training set of images into spatially organised multi-class SVM models (using BOW and Edge features) and re-classifed the test collection using the hybrid approach combining SVMs and the hierarchical approach. Results of this investigation are again shown in Table 1. Of the 270 images within the test collection, the BOW SVMs recognised 156 of them while the Edge based SVMs recognised 214 of the test images. The Edge based hybrid approach slightly outperforms the BOW using this dataset.

Tagging. The selection of appropriate tags for the target image is extremely important within the workflow since tags are used for image annotation and as the input to the web-based augmentation stage. In order to evaluate the accuracy of the tagging phase, we groundtruthed 85 test images. The tags for the images returned for each image by the image matching service were formed into a union pool set. An annotator manually judged each pooled tag and made a binary classification of its relevance (ie. relevant or not). A tag was determined to be relevant if it described the landmark featured in the image. As such associated tags were not deemed relevant. Thus details such as descriptions of camera types, related tags such as weather or lighting, or information on the year or events and activities were deemed non-relevant, since the emphasis in the groundtruthing was placed on tags which identify the landmark in the image. On average 9.85 tags were deemed relevant per image, while each image had an average of 69.09 tags taken from 12.85 images matches from the image collection.

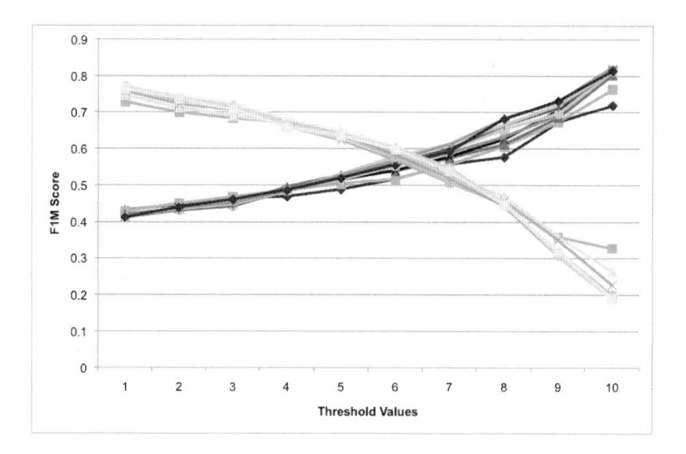

Fig. 6. The precision can be seen to increase for each of the weighting approaches as the thresholds increase, while the recall increases

With a groundtruth established, the tag weighting and thresholding approach as previously described was applied to the image matching results for each test image. The input parameters were varied from 0.5-0.95 for both weighting and threshold and all of the combinations iterated. The set of tags returned from each variation was compared against the groundtruth for that image. Precision and recall measures were calculated as outlined in [8]. These were then averaged across all of the test images and the F1M measure calculated.

Within the selection of tags there was a need to balance precision and the recall so that 'noisy' or superfluous tags are kept to a minimum while a maximum of the desired tags are contained within the selected set. Applying a low threshold results in an unconstrained and highly noisy set displaying high recall but extremely low precision. This is illustrated in Fig 6. Conversely, a higher threshold and weight results in an overly constrained set, which while displaying high precision, has very low overall recall. By exploring the various variations for the highest F1M, we identified a threshold value of 0.75 with an iteratively decreasing tag weighting of 0.85 to be optimal for tag selection. This resulted in on average 13.9 tags being selected. While some of the selected tags are not directly relevant, this may be acceptable to users since many of these additional tags are noted to be ancillary or related descriptors rather than genuine noise.

Toponym identification. A toponym is used to initiate the fact extraction and title augmentation step, and its accuracy is thus important to the effectiveness of the fact extraction stage. The identification of the appropriate toponym label for each image is reliant upon the outputs of the tag filtering and selection process as outlined previously. To investigate this, 87 test images were selected, image matching was performed and a set of tags filtered from the results selected. Nearby toponyms were looked up using GeoNames and a candidate selected based on the tag set in a manner as outlined in Section 3.2. Each of the returned toponyms was then annotated into one of the following categories: Incorrect toponym identified; Vague or unspecific toponym identified, e.g. Paris, France; Toponym is related to the target but is incorrect, this included a landmark nearby or within the image but which was not the primary focus or featured landmark, e.g. the Champ de Mars returned in place of the Eiffel Tower; and finally a correctly identified toponym. In total 13 of the toponyms were incorrect, 4 were vague, 10 were incorrect but related and 60 were correct.

While 68.97% of the tested images returned a correctly identified toponym, a further 16% (vague and related categories) may be considered acceptable (totalling 88.5%). All of the vague cases were composed of a generic toponym of 'Paris', which returned facts such as 'Paris is named after a Celtic tribe called the Parisii who lived on the island in the river', 'Paris is famous for its huge number of cafes and brasseries' and 'Paris was made for lovers and lovers of life'. While these facts are not ideal, they are generic enough to be reasonably acceptable. Additionally those which are related often contained reference to the target landmark. For example, in the case where the Champ de Mars was identified in place of the Eiffel tower, the first returned fact is the following: 'Champ de Mars is a green area located in the middle of the Eiffel Tower and the Ecole Militare building'. In another case, where the Champs Elysees was returned instead of the Arc de Triomphe, the first fact returned again referenced the desired landmark: 'Champs-elysees is a seventeenth century garden-promenade turned avenue connecting the Concorde and Arc de Triomphe'.

Fact Extraction and Title Augmentation. For this evaluation 68 place names from around Europe were selected. We chose an even mixture of urban / rural and famous / not famous places from European cities (London, Riga, Zurich and Dublin) and countryside (UK, Latvia, Switzerland and Ireland), and various types of place - churches, statues, mountains, rivers, etc. For each place the top ranked fact was evaluated in terms of its correctness (according to one investigator consulting relevant websites) and whether or not it was deemed interesting (according to the judgments of five subjects). Our evaluation criteria were actually rather strict, since it was found that a majority of subjects rated more facts as 'interesting' (78%) than we ourselves rated as correct (50%).

Overall we are encouraged by the performance of each of the component technologies used in our application. While all of them use some degree of empirical design and parameter selection, none of these specifically focus on the dataset used here and they can easily be adjusted for other similar environments. We anticipate that in an operational application empirical parameters could be adjusted automatically based on user feedback.

5 Conclusions and Further Work

This paper has described our novel integrated system for automatically captioning landmark images captured on a GPS enabled mobile device. Evaluation of the three principle components of the systems shows that they each have a high degree of effectiveness. They do however make some mistakes. However, informal testing of the combined system shows that the application provides good tags for popular landmarks for which there are many existing labeled images contained in online social collections with a correspondingly large number of facts available online. Tagging is less effective for less frequented landmarks where there are less online images and less web content available. For this latter case we plan to explore more sophisticated techniques to improve the quality of tagging, however it is likely that for many less popular landmarks this problem will address itself over time as more online content appears naturally. We also plan to carry out an end-to-end evaluation of the captioning system. This will evaluate the accuracy of landmark identification, factual augmentation, and the acceptability and value of the captions to users.

In the longer term, users will be able to upload captioned images to social image collections. Over time this will expand the number and detail of annotated images available. This itself will provide more effective sources of information for landmark identification and for the selection of accurate and interesting tags. While the system described here is only implemented for English, the methods used are all language independent and it could easily be ported to other languages by means of new stopword lists, localised toponyms lists and collections of word patterns in the fact extraction stage [16].

Acknowledgement. The research reported in this paper is part of the project TRIPOD supported by the European commission under contract No. 045335.

References

1. Geonames, `http://www.geonames.org`
2. Panoramio, `http://www.panoramio.com`
3. Yahoo! search boss, `http://developer.yahoo.com/search/boss/`
4. Bay, H., Tuytelaars, T., Van Gool, L.: SURF: Speeded up robust features. In: Leonardis, A., Bischof, H., Pinz, A. (eds.) ECCV 2006. LNCS, vol. 3951, pp. 404–417. Springer, Heidelberg (2006)
5. Chevallet, J.-P., Lim, J.-H., Leong, M.-K.: Object identification and retrieval from efficient image matching. Snap2Tell with the STOIC dataset. Information Processing and Management 43(2), 515–530 (2007)
6. Cortes, C., Vapnik, V.: Support-vector networks, vol. (3), pp. 273–297 (1995)
7. Fritz, G., Seifert, C., Paletta, L.: A mobile vision system for urban detection with informative local descriptors. In: Proceedings of the IEEE International Conference on Computer Vision Systems (ICVS 2006), p. 30 (2006)
8. Jäschke, R., Eisterlehner, F., Hotho, A., Stumme, G.: Testing and evaluating tag recommenders in a live system. In: Workshop on Knowledge Discovery, Data Mining, and Machine Learning, pp. 44–51 (2009)
9. Lorenz Wendt, F., Bres, S., Tellez, B., Laurini, R.: Markerless outdoor localisation based on sift descriptors for mobile applications. In: Elmoataz, A., Lezoray, O., Nouboud, F., Mammass, D. (eds.) ICISP 2008. LNCS, vol. 5099, pp. 439–446. Springer, Heidelberg (2008)
10. Lowe, D.G.: Local feature view clustering for 3D object recognition. In: Proceedings of the 2001 IEEE Computer Society Conference on Computer Vision and Pattern Recognition, CVPR 2001, vol. 1, pp. I-682–I-688 (2001)

11. Lowe, D.G.: Distinctive image features from scale-invariant keypoints. International Journal of Computer Vision 60, 91–110 (2004)
12. Malobabic, J., le Borgne, H., Murphy, N., O'Connor, N.: Detecting the presence of large buildings in natural images. In: Proceedings of the 4th International Workshop on Content-Based Multimedia Indexing (CBMI 2005), pp. 529–532 (2005)
13. Porter, M.F.: An Algorithm for Suffix Stripping. Program 14(3), 130–137 (1980)
14. Qingji, G., Juan, L., Guoqing, Y.: Vision based road crossing scene recognition for robot localization. In: Proceedings of the International Conference on Computer Science and Software Engineering, vol. 6, pp. 62–66 (2008)
15. Rahmani, R., Goldman, S.A., Zhang, H., Cholleti, S.R., Fritts, J.E.: Localized content-based image retrieval. IEEE Transactions on Pattern Analysis and Machine Intelligence 30(11), 1902–1912 (2008)
16. Salway, A., Kelly, L., Skadina, I., Jones, G.J.F.: Portable extraction of partially structured facts from the web. In: Loftsson, H., Rögnvaldsson, E., Helgadóttir, S. (eds.) IceTAL 2010. LNCS, vol. 6233, pp. 345–356. Springer, Heidelberg (2010)
17. Szummer, M., Picard, R.W.: Indoor-outdoor image classification. In: Proceedings of the IEEE International Workshop on Content-Based Access of Image and Video Database, pp. 42–51 (1998)
18. van Rijsbergen, C.: Information Retrieval, 2nd edn., Butterworths (1979)
19. Yeh, T., Tollmar, K., Darrell, T.: Searching the web with mobile images for location recognition. In: Proceedings of the 2004 IEEE Computer Society Conference on Computer Vision and Pattern Recognition (CVPR 2004), vol. 2, pp. 76–81 (2004)

IROM: Information Retrieval-Based Ontology Matching

Hatem Mousselly-Sergieh[1] and Rainer Unland[2]

[1] Chair of Distributed Information Systems, University of Passau,
Passau, Germany
[2] Data Management Systems and Knowledge Representation Group,
University of Duisburg-Essen,
Essen, Germany

Abstract. A crucial piece of semantic web development is the creation of viable ontology matching approaches to ensure interoperability in a wide range of applications such as information integration and semantic multimedia. In this paper, a new approach for ontology matching called IROM (Information Retrieval-based Ontology Matching) is presented. This approach derives the different components of an information retrieval (IR) framework based on the information provided by the input ontologies and supported by ontology similarity measures. Subsequently, a retrieval algorithm is applied to determine the correspondences between the matched ontologies. IROM was tested with ontology pairs taken from two resources for reference ontologies, OAEI and FOAM. The evaluation shows that IROM is competitive with top-ranked matchers on the benchmark test at OAEI campaign of 2009.

Keywords: Ontology matching, information retrieval, ontology similarity.

1 Introduction

Ontologies were proposed as a means of knowledge sharing. They have found a wide acceptance and usage in various application domains. However, developing a universal ontology that covers the different aspects of some domain of interest is impractical. In fact, it is common to have more than one ontology modeling the same domain from different perspectives and at different levels of granularity. On the other hand, combining or merging ontologies is considered beneficial for achieving interoperability. Therefore, getting the best benefit from using ontologies depends in the first place on providing mechanisms for the discovery of potential overlaps.

In the filed of semantic multimedia, for instance, several ontologies have been proposed to bridge the semantic gap between low level features of multimedia object content and its high level conceptual meaning. Nevertheless, these ontologies are often designed for a particular application. Therefore, finding the correspondences between these heterogeneous ontologies would contribute to solving

T. Declerck et al. (Eds.): SAMT 2010, LNCS 6725, pp. 127–142, 2011.

several interoperability issues that coexist in semantic multimedia annotation and retrieval.

Ontology matching is a field of research that deals with this problem. It provides approaches for discovering a set of *correspondences*, called *alignment* among ontologies. In recent yeas, several ontology matching approaches have been proposed. However, investigating the role of applying IR on the matching problem is still limited. To our best knowledge PRIOR+ [12] is the only approach that explicitly mentions the usage of IR.

In this paper, we combine IR and ontology similarity measures and propose a new structure-based ontology matching approach called IROM. IROM parses a pair of input ontologies and generates for each entity (in both ontologies) a set of terms called entity descriptions. A part of the description terms are derived from entity's metadata like their labels and identifiers, while the other part is extracted by applying ontology similarity measures. The generated entity descriptions are then gathered in similarity matrices (one for each ontology). The resulted matrices provide the required elements for mapping the ontology matching problem to information retrieval models. Subsequently, an IR algorithm is applied and a pre-alignment for the ontology pair is generated. The pre-alignment is then filtered to produce the final alignment.

The rest of the paper is organized as follows: In the next section the related work is presented. In section 3 the IROM algorithm is introduced with background information on the applied techniques. In section 4 IROM is evaluated and in section 5 we conclude and discuss our future work.

2 Related Work

Ontology matching has attracted the attention of many researchers in recent years. [4] provides an intensive survey of different ontology matching approaches. In this paper we only consider the 5 top matchers that have participated at the benchmark test track of the Ontology Alignment Evaluation Initiative[1] (OAEI) of the year 2009. These matchers represent the recent advances in the ontology matching domain and apply diverse techniques. Additionally, we consider PRIOR+ since it also applies IR during its matching process.

Anchor-Flood [15] starts by selecting a pair of concepts called the *anchor*. For each anchor concept the matcher collects small blocks of neighboring concepts based on their ontological links. The concepts in the blocks are then aligned based on the lexical information and the structural relations. In turn, the aligned concepts are dealt with as new anchors. The process repeats until there is no anchor left.

Lily [18] extracts *semantic subgraphs* from the matched ontologies and generates initial alignment based on the linguistic as well as the structural similarity of the semantic subgraphs. Furthermore, a propagation strategy can be applied to find additional correspondences. A similarity matrix is created, the entries of

[1] `http://oaei.ontologymatching.org/`

which represent the confidence level between the entities of the matched ontologies. The produced alignment is then filtered by removing all correspondences which have confidence levels under a certain threshold. The threshold is determined by applying an image threshold selection algorithm which is based on the maximum entropy.

DSSim [13]: considers the entities of one ontology equivalent to user queries while the entities of the other as a set of possible answers. DDSim applies different kinds of similarity measures named *experts* on the ontological entities and then uses the Dempster-Shafer theory to determine the degree of similarity between the entities.

RiMOM [17] is based on the Bayesian decision theory. It applies different kinds of strategies called *decisions* to discover the correspondences between the ontologies. The decisions are selected according to the nature of the matched ontologies. The set of the decisions include: Name-, instance-, description-, taxonomy context- and constraints-based decisions. To improve the decision RiMOM also applies natural language processing (NLP) techniques. The alignments of the selected decisions are then combined and the produced alignment is refined heuristically.

ASMOV [6] is an iterative process which is split in two main steps: Similarity calculation and semantic verification. The similarity calculation step applies different kinds of similarity measures (lexical, structural and extensional) on the ontological entities and produces a similarity matrix. A pre-alignment is extracted from the similarity matrix by selecting the entity pairs with the maximum similarity values. A semantic verification step is then applied to eliminate the correspondences that contradict with the semantic of the matched ontologies. The process repeats until a finalization condition is satisfied.

PRIOR+ [12] generates different kinds of similarity matrices (name, structure and profile) by applying the vector space model. For each kind of similarity a weight is produced by applying a specialized measure, called the *harmony* measure, on the corresponding similarity matrices. The similarity values are then combined to determine the final similarity value between the entities of the of the matched ontologies. PRIOR+ is also able to apply an interactive and competition network (IAC) to find a solution that best satisfies the constraints of the ontology.

It is important to note that all presented approaches rely on using different kind of similarity measures (string-based, structure-based, etc.). In contrast, IROM uses only one string-based similarity which is the edit-distance [11]. Moreover, the term similarity matrix in IROM refers to a matrix containing similarity values for entity pairs of the same ontology, while in the other approaches the similarity matrix represents the degree of correspondences between entities of the matched ontologies.

DSSim and IROM share the idea of considering one ontology as a source for user queries and the other for document collection, however, they apply different processes.

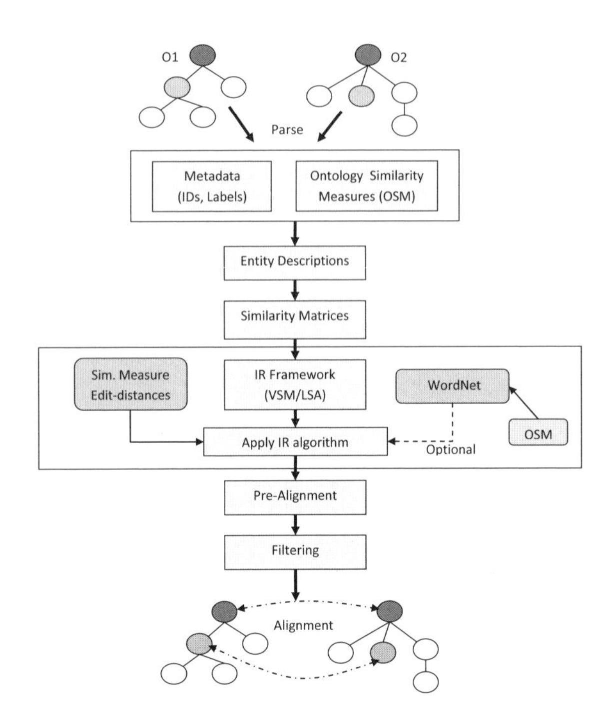

Fig. 1. The architecture of IROM

On the other hand, entity descriptions in IROM correspond to both profile and structural similarities of PRIOR+. Furthermore, both matchers (IROM and PRIOR+) apply IR, however, IROM provides two variants of the vector space model while PRIOR+ applies only the simple vector model. Additionally, IROM applies a special index-terms weighting approach which is based on ontology similarity measures (see the next section), while PRIOR+ uses the traditional approach of term frequency-inverse document frequency (tf.idf) [1].

3 The Matching Algorithm

Figure 1 shows the architecture of IROM. Before we discuss the way IROM works, we briefly introduce the applied IR models as well as ontology similarities which are considered the basic building blocks of IROM approach.

3.1 Information Retrieval

IROM has been equipped with two variants of the vector space models. This allows us to compare the different models and study their effects on different ontology matching scenarios. The provided IR models are the traditional vector space model (VSM) [1] and the latent semantic analysis (LSA) [9].

In IR the term *query* refers to a user request to find a certain piece of data. The set from which the answers for a query are collected is called the *document* set.

In VSM a vector space is defined based on a set of terms extracted from the document collection, called the *index terms*. Queries and documents are represented as vectors and *weights* are used to determine the degree of relevance between the index terms and the corresponding documents/queries. To determine the similarity between a document-query vector pair the cosine of the angle between the two vectors is calculated and used as a rank to the similarity between them.

Obviously, the similarity between the document and the query vectors in VSM depends essentially on the set of their syntactically equivalent index terms. However, it is possible for a document to match a query even if they do not share any syntactically similar index terms as in the case of using synonyms. On the other hand, queries and documents sharing some homonym index terms should be recognized as dissimilar.

LSA - a variant of vector space model deals with such situations. The main idea of LSA is to perform a retrieval in a reduced vector space that better represents the semantic implied by the original space. LSA starts by building a $t \times d$ matrix called the term-document matrix. The entries of that matrix hold weights representing the degree of relevance between the term-document pairs. The resulted term-document matrix is then divided into a multiplication of three matrices (normally using Singular Value Decomposition (SVD)). This decomposition allows us to choose an approximation to the term-document matrix that better captures the hidden semantic. The document and the query vectors are then represented in the new reduced space and their similarity can again be calculated by applying the cosine measure.

3.2 Ontology Similarity

The other essential component of IROM is the ontology similarity measures. IROM allows the use of different kinds of measures. This enables us to investigate the effect of different ontology similarity approaches on the matching quality.

Ontology similarity [14] is defined as a measure to the degree of relatedness between the concepts of the same ontology. Two concepts are considered to be related if they are connected via any kind of relation (e.g. IS-A).

We use ontology similarity measures to provide the weights for the term-document/term-query pairs. These measures generate for each concept in the ontology a set describing concepts of different degrees of relatedness. For example, in Figure 2 the concepts *Thing* and *Vehicle* are connected to the concept *Automobile*, thus, they can be used to describe it. However, it is obvious that *Thing* is less similar to *Automobile* than *Vehicle* is. Therefore, when doing a structural comparison between *Automobile* and a concept from another ontology, *Thing* must play a less important role in determining the similarity than *Vehicle* should do. Considering that, we have defined the following requirements on the ontology similarity measure which will be used as a weighting system for the applied IR models:

- The similarity measure should only consider the information provided by the ontology itself. Therefore, we do not consider similarity measures that use external resources such as corpus-based measures (e.g. Jiang & Conrath [7]).The reason for that is to avoid additional computation.
- The similarity measure should satisfy the depth property [16] which implies that the distance represented by an edge in the ontology should be reduced with an increasing depth of the location of the edge. This property ensures that the deeper the concepts in the ontology the more similar they are.

There are a number of ontology similarity measures that satisfy these properties (For a survey of the different ontology similarity measures see [2, 16]). IROM allows us to choose between two measures: The Conceptual Similarity (WP) [19] measure and the Weighted Shared Nodes (WSN) [16] measure.

WP is path-based measures that defines the similarity between two concepts in the ontology based on the length of the path between them. In order to satisfy the depth property WP scales the calculated path length by the distance between the least common super-concept of both concepts and the root of the taxonomy. WP is given in the following formula:

$$Sim_{wp}(c_1, c_2) = \frac{2 \times depth(c_3)}{N_1 + N_2 + 2 \times depth(c_3)}$$

Where:

- c_1, c_2 are two concepts in the ontology,
- c_3 is the least common super concept of c_1 and c_2,
- N_1 is the number of **links** in the path from c_1 to c_3,
- N_2 is the number of **links** in the path from c_2 to c_3,
- depth(c_3) is the number of **nodes** in the path from c_3 to the root concept.

In contrast to WP, which is only based on IS-A relation, WSN uses every kind of semantic relation to determine the similarity between concepts, thus, it is called a multi-path measure. Additionally, WSN assigns for each concept in the ontology a weight according to the semantic relation by which they are reached. The concepts which are reachable from a given concept are gathered in a fuzzy set with their weights indicating their corresponding degree of membership. Another difference between WSN and WP is that WSN satisfies the generalization property defined by [16] which indicates that *"concept inclusion (IS-A) implies reduced similarity in the direction of inclusion"*. That means, WSN is an anti-symmetric measure. For example, in the ontology of Figure 2 WSN considers the similarity between *Thing* and *Vehicle* in the direction from *Thing* to *Vehicle* to be greater than their similarity in the opposite direction, i.e., form *Vehicle* to *Thing*. Mathematically, WSN is defined as follows:

$$Sim_{wsn}(c_1, c_2) = \rho \times \frac{|\alpha(c_1) \cap \alpha(c_2)|}{|\alpha(c_1)|} + (1 - \rho) \times \frac{|\alpha(c_1) \cap \alpha(c_2)|}{|\alpha(c_2)|}$$

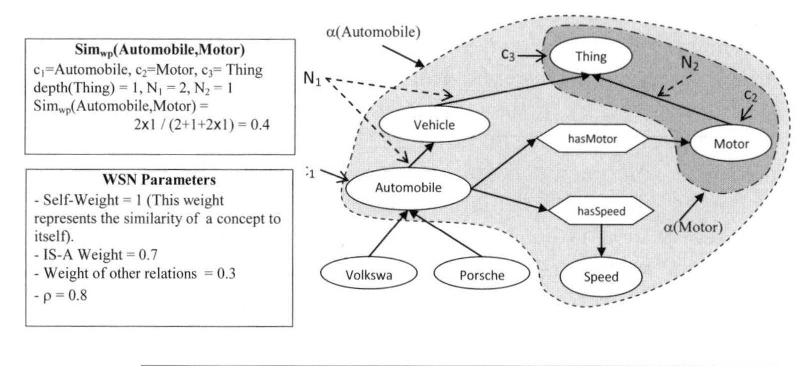

Sim$_{wsn}$(Automobile,Motor)
|α(Automobile)| = |{1/Automobile + 0.7/Vehicle + 0.7*0.7/Thing + 0.3/Motor+ 0.3/Speed}| = 2.79
|α(Motor)| = |{1/Motor+ 0.7/Thing }| = 1.7
|α(Automobile ∩ Motor)|= |{0.3/Motor+ 0.49/Thing }| = 0.79
Sim$_{wsn}$(Automobile, Motor) = 0.8 * (0.79/2.79) + (1-0.8) * (0.79/1.7) = 0.319

Fig. 2. Ontology similarity measures (WSN and WP) applied on sample ontology (Modified from [3])

Where:

- c_1, c_2 are two concepts in the ontology.
- $\rho \in [0,1]$ determines the degree of the influence of the IS-A relation (for a discussion see [16]).
- $\alpha(c_1)$ is the set of nodes reachable from c_1.
- $\alpha(c_2)$ is the set of nodes reachable from c_2.
- $\alpha(c_1) \cap \alpha(c_2)$ is the set of nodes reachable by c_1 and c_2 at the same time (shared nodes).

Figure 2 gives sample calculations of both measures.

3.3 How It Works

IROM matches ontology pairs written in the Web Ontology Language (OWL) based on the information provided by the ontology structure only (intensional dimension). It does not consider information which are provided by ontology instances (extensional dimension). Furthermore, no reasoning is applied on the ontologies before they get matched.

Concepts and roles of the ontology constitute the matching entities of IROM. For each entity of each ontology an entity description is generated. According to the type of the entity, we recognize two kinds of entity descriptions.

Concept Descriptions: Include terms derived from the concept's metadata such as identifiers and labels (OWL elements rdf:ID and rdfs:label respectively). The descriptions are also enriched with additional terms that are extracted by applying ontology similarity measures on ontology concepts. This results in identifying the set of concepts that are closely related [16]. Each of those concepts can then be used to describe the other concepts in the set.

In cases where the concept is defined as a combination of other concepts (in OWL the elements owl:Union and owl:Intersection), description terms are derived from the component concepts. Here we recognize two cases: The union case, in which the similarity of the combined concept to its components is given a constant value (w_1), while the similarity, in the opposite direction, i.e, from the components to the combined concept, is considered proportional to the number of the components (n) and is calculated as ($w_2 = w_1/n$). Intersection represents the reverse case, component concepts are considered equally similar to the combined concept (the weight w_1), whereas the similarity between the combined concept (intersection concept) and its components is considered proportional to the number of the components (the weight $w_2 = w_1/n$).

Role Descriptions: Ontology roles correspond to the elements owl:ObjectProperty and owl:DatatypeProperty of OWL. Like in concepts' case, identifier and label elements are used as description terms. Additionally, domains and ranges of a property are used to enrich role descriptions with additional terms. Concepts representing domains and ranges of a role are given predetermined weights representing their relatedness to their corresponding role.

Similarity Matrix: We combine the entity descriptions of the ontology in a single component called the similarity matrix. It is an $n \times m$ matrix whose rows are the set of all description terms of all generated entity descriptions and its columns are the entities of the ontology. The entries of this matrix represent the similarity values/weights of the term-entity pairs. The main advantage of the similarity matrix is that it contains the elements needed by the applied IR models, i.e., (1) the index terms, which correspond to the rows of the matrix, (2) the documents/queries, which correspond to the columns, and (3) the corresponding term-document weights which correspond to the entries of the matrix. Furthermore, the similarity matrix can easily be extended with terms extracted from additional ontology language features such as comments.

Applying VSM: At first, we assume that one of the input ontologies is a source for document collection while the other one is a source for queries. The t-dimensional space S^t of VSM for ontology matching is then defined as follows: Let QT be the set of all description terms of all entity descriptions of the first ontology and DT be the set of all description terms of all entity descriptions of the second ontology: $S^t = QT \cup DT$ with $t = |QT| + |DT|$. Queries as well as documents (which correspond to ontology entities) are represented as vectors over S. The values of the vector components correspond to the similarity values of the corresponding description terms.

To check a query-document pair for similarity, the cosine measure is applied. During the computation, the components of the two vectors are compared syntactically using the edit-distance measure. If two index terms of a query and a document vector, respectively, are syntactically equivalent, their weights stay intact. Otherwise the weights are modified according to the calculated distance.

The query-document pair is considered to match each other if the cosine value exceeds a predetermined value called the ranking threshold.

Applying LSA: To enable the use of LSA in the context of ontology matching a suitable LSA space [9] has to be defined. An LSA space has three dimensions: (1) The corpus from which the index terms are generated, (2) the terms-document weighting system, and (3) the optimal number of dimension that the term-document matrix should be reduced to.

The similarity matrix covers the first two dimensions of the required LSA space. In fact, the rows of the similarity matrix correspond to the terms of the term-document matrix, while its columns correspond to the documents. Additionally, the entries of the similarity matrix can be used as weights for the term-document pairs. Here, we use the similarity matrix corresponding to one of the input ontologies to be a source for LSA components, while the other similarity matrix is used to provide the queries.

Determining an optimal dimension for LSA remains elusive and there is no optimal procedure yet. It is argued that is because optimal dimensionality is task, content and size dependent (Chapter.4 of [9]). On the other hand, [10] suggest that using an LSA dimension between 100 and 300 is expected to provide good retrieval.

In this work we took an experimental approach to determine an optimal dimensionality of the LSA space for ontology matching. According to [8] (Section 4) the performance of LSA approaches that of traditional vector space retrieval when the number of LSA dimensions approaches the rank of the term-document matrix. Therefore, we decided to consider the rank of the term-document matrix as an upper bound for the number of LSA dimensions.

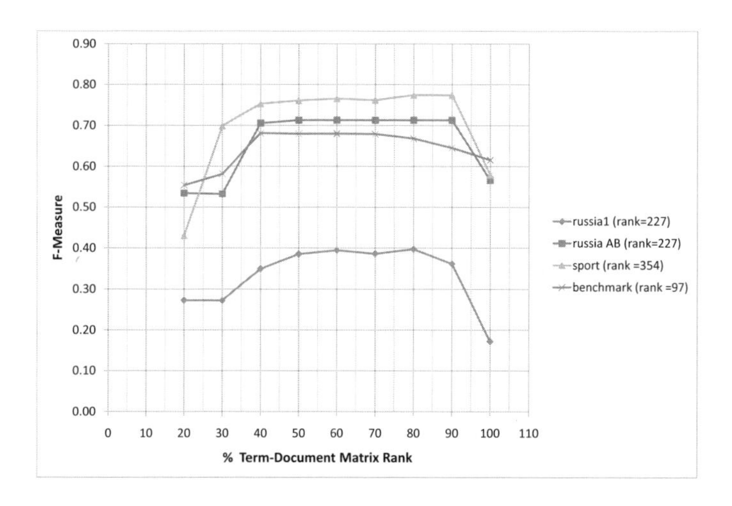

Fig. 3. Ontology pairs matched by applying LSA to determine an upper bound for the optimal number of dimensions

On the other hand, determining a lower bound is done as follows: We used IROM to match ontology pairs of different natures (taken from OAEI and FOAM[2]). We started by an LSA dimension equals 20% of the rank of term-document matrix (R) and increased it by the factor of 10 to reach the full value of R (see Figure 3). The analysis shows that the f-measure increases with an increasing number of dimensions and reaches its best value when the number of dimensions reach 40 to 50% of R. After that, LSA shows a constant performance until the number of dimensions approaches R, then the performance starts to decrease. Based on this evaluation we found that choosing a number of dimensions between 45 to 65% of R would provide best matching results.

After LSA dimensions have been determined, document and query vectors are represented in the new reduced space. In the same way as in VSM, the similarity between document and query vectors is determined by the cosine measure.

Filtering the Pre-Alignment: In IR for each query a ranked set of related documents is delivered. However, in ontology matching, only correspondences of best ranks are of interest. On the other hand, using ontology similarity measures as a weighting system makes some of the correct correspondences implicit. This can be explained by the following example.

Considering the ontology pair and the matching configuration of Figure 4. IROM produces the pre-alignment shown in Table 1 (it is actually the set of all possible correspondences).

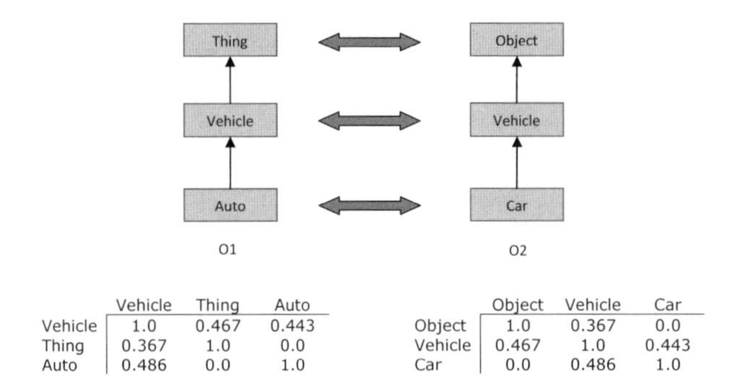

	Vehicle	Thing	Auto
Vehicle	1.0	0.467	0.443
Thing	0.367	1.0	0.0
Auto	0.486	0.0	1.0

	Object	Vehicle	Car
Object	1.0	0.367	0.0
Vehicle	0.467	1.0	0.443
Car	0.0	0.486	1.0

Fig. 4. O1 and O2 with their corresponding similarity matrices. WSN is applied (IS-A Weight = 0.5, Semantic Relation Weight = 0.1 and Similarity Threshold = 0.3). The arrows between concept pairs indicate that the two concepts should be aligned together (reference alignment).

As we can see, all correspondences of the reference alignment are included in the result set. However, only the two *Vehicle* concepts were classified as a best match, while the other reference correspondences were not recognized as such. For example *Car* is recognized to better much *Vehicle* than *Auto*. Such a

[2] http://www.aifb.uni-karlsruhe.de/WBS/meh/foam/

Table 1. Pre-alignment of O1 and O2 (Ranking Threshold = 0.1)

O2	O1	Rank
Car	Vehicle	0.346
Car	Thing	0.171
Car	Auto	0.164
Object	Vehicle	0.361
Object	Thing	0.179
Object	Auto	0.171
Vehicle	Vehicle	0.729
Vehicle	Thing	0.361
Vehicle	Auto	0.346

situation can be explained based on the vector representations of these concepts. *Car* can be described as (Vehicle:0.443, Car:1), *Vehicle* (of O1) as (Vehicle:1, Thing:0.376, Auto:0.486) and *Auto* as (Vehicle:0.443, Auto:1). Since *Vehicle* is the only shared component among the vectors corresponding to these concepts the cosine similarity between *Car* and *Vehicle* as well as *Car* and *Auto* will depend on the multiplication of the corresponding weights of the *Vehicle* component. The greater the value of the multiplication the more similar are the concepts. For the *Car-Vehicle* pair this multiplication is $0.433 \times 1 = 0.443$ while for the *Car-Auto* pair it is $0.433 \times 0.443 = 0.196 < 0.443$. Therefore, *Car* is classified to be more similar to *Vehicle* than to *Auto*.

To handle such a situation, a filtering algorithm is proposed(see Listing 1.1). It refines the calculated pre-alignment by removing superfluous correspondences and keeping the most relevant ones.

Let O_1 and O_2 be two ontologies, a correspondence is a tuple $a(e_1, e_2, r)$ where $e_1 \in O_1$, $e_2 \in O_2$ are ontology entities and $r \in [0, 1]$ is a rank (corresponding to the degree of confidence [5]). The pre-alignment I is the set of found correspondences. Initially, the algorithm sorts the pre-alignment I according to the ranks of the correspondences in descending order. Then it starts with the first correspondence a, adds it to the final alignment FN and removes all other correspondences b in which either of both entities of a appears. This process is repeated until there is no more correspondences to be handled. Eventually, the refined set will contain for each entity only one correspondence with the highest rank which exceeds the predefined ranking threshold.

```
FN = Φ //The final alignment
I := Sort_rank(I) //sort according to correspondences rank
For each a ∈ I
        FN := FN ∪ {a}
    For each b ∈ I − {a}
            IF b.e1 ∈ {a.e1, a.e2} ∨ b.e2 ∈ {a.e1, a.e2}
                    I := I − {b}
```

Listing 1.1. The filtering algorithm

We will apply the algorithm on the pre-alignment of Table 1. The correspondence (*Vehicle*, *Vehicle*, 0.729) has the best rank, so it is added to the final align-

Table 2. Step 1 of the filtering algorithm

O2	O1	Rank
Car	Auto	0.164
Object	Thing	0.179
Object	Auto	0.171
Vehicle	Vehicle	0.729

Table 3. Step 2 of the filtering algorithm

O2	O1	Rank
Car	Auto	0.164
Object	Thing	0.179
Vehicle	Vehicle	0.729

ment. Furthermore, correspondences containing *Vehicle* as a matched entity are removed from the list (Table 2). After the correspondence (*Object*, *Vehicle*, 0.361) has been removed, the correspondence (*Object*, *Thing*, 0.179) has the best rank. So, it is added to the final alignment and all correspondences containing *Object* or *Thing* are removed. This elimination results in adding the correspondence (*Car*, *Auto*, 0.164) to the final alignment (Table 3) and finishing the process.

4 Evaluation

A prototypical implementation of IROM that aligns ontologies written in OWL version 1.1 was used to evaluate the correctness and the performance of IROM.

For this purpose, we used test data taken from OAEI, the campaign of year 2009, and FOAM. The benchmark track of OAEI provides test data that cover a wide range of ontology language features. While FOAM provides ontologies of different sizes and natures (we took 8 ontology pairs out of the provided ontology pairs).

We compared the alignments produced by IROM with those of the 5 top matchers which performed best at the benchmark track of OAEI (campaign 2009)[3]. The evaluation is based on the standard measures applied by OAEI: Precision, recall and f-measure.

The evaluation is done to answer the following questions: (1) How well does IROM perform in comparison to the other top matchers which have participated at OAEI campaign 2009? (2) How do different ontology similarity measures affect the matching performance? (3) What is the effect of different IR models on the matching performance and which models fit better to which cases?

IROM vs. Other Matchers. Two configurations of IROM, we call them IROM I and IROM V respectively, were evaluated against the benchmark track and the FOAM reference ontologies. IROM I uses WP while IROM V uses WSN and both matchers apply VSM retrieval. The performance of both IROM configurations was compared to that of the five top matchers at OAEI 2009 benchmark.

Results for Benchmark of OAEI 2009:
Benchmark data set is divided into three groups: 1xx, 2xx, and 3xx. The group 2xx in turn is further divided into three subgroups (201-210, 221-247 248-266).

[3] Results can be found under: `http://oaei.ontologymatching.org/2009/results/benchmarks.html`

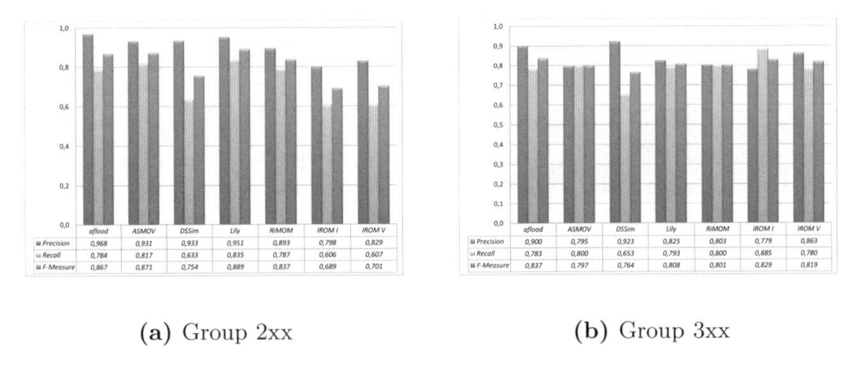

(a) Group 2xx **(b)** Group 3xx

Fig. 5. Precision, recall and f-measure produced by different matcher for ontology pairs of benchmark dataset. (aflood corresponds to Anchor-Flood)

For the first group (1xx), and like all other participating matchers, IROM (I and V) was able to find the complete set of reference correspondences and so producing an f-measure equals to 1.

For the group 2xx, IROM (I and V) provided a very good matching performance for the second subgroup (221-247) with an f-measure value of 0.988 and did also relatively well for the first subgroup (201-210) providing an f-measure value of 0.702. However, IROM (I and V) provided a less good performance for the third subgroup. This can be explained based on the observation that ontology pairs belonging to this subgroup share a very small set of syntactically similar concepts and roles (the same applies to some ontology pairs of the first subgroup). Instead, they only share individuals. Therefore, producing good matching results for this subgroup relies on using the information implied by ontology instances, the thing that IROM does not support yet. The overall performance of IROM for group 2xx is depicted in Figure 5a.

In contrast to the previous group, IROM (I and V) performed very well when matching ontology pairs of group 3xx. IROM came second after Anchor-Flood with a very slight difference (Figure 5b) and outperformed the other matchers.

Results for FOAM Ontologies:
To emphasize the ability of IROM to compete with other matchers and to provide good matching results, we compared the performance of IROM to that of Lily and Anchor-Flood (Lily came first and Anchor-Flood came second at the benchmark of OAEI 2009). We chose 8 ontologies pairs of different natures and sizes to perform this test.

Out of the eight ontology pairs, IROM I provided best f-measures for 4 of them (animals, Russia AB, Russia CD, and Network). On the other hand, IROM V performed best for 3 pairs (animals, People and Pets and Network). Furthermore, the results produced by both IROM configurations (I and V) for the remaining pairs were not far from the best results (Figure 6a).

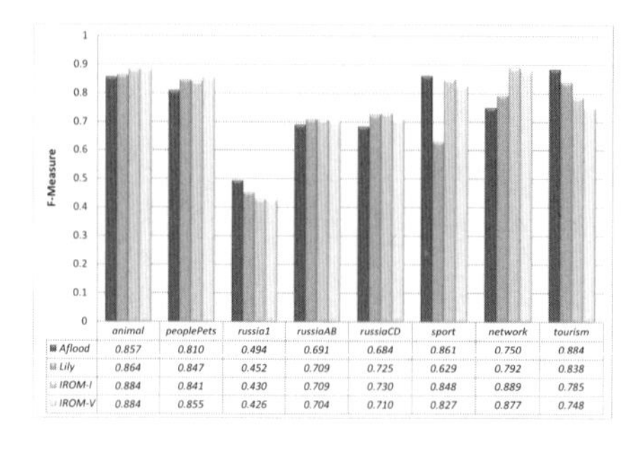

(a) F-measure

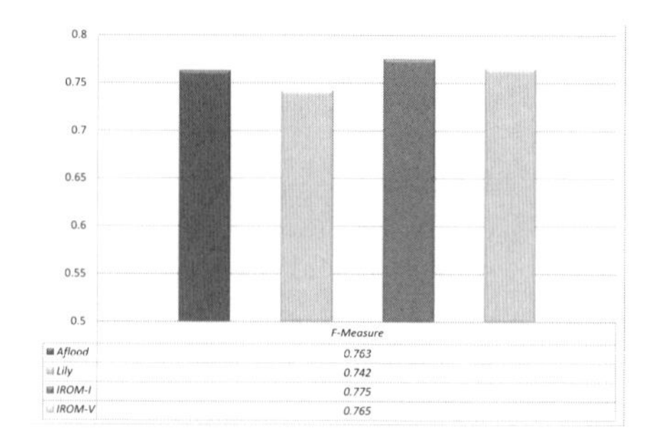

(b) Average f-measure

Fig. 6. Comparing Anchor-Flood, Lily and IROM (I and V) according to FOAM ontologies.

In total, IROM provided the best results for the selected FOAM ontologies. This is shown through the values of the average f-measures[4] produced by the four matchers (Figure 6b).

Different Ontology Similarity Measures. IROM I (uses WP) and IROM V (uses WSN) showed a very close matching performance. However, deciding which similarity measure to use depends on the degree of the syntactic overlap between the two ontologies. This can be explained by the following example. Assume that we have two concepts both labeled as "A" (syntactically identical) and belong to two different ontologies. The two concepts are root concepts and linked via properties "hasC" and "hasD" to the concepts "C" and "D", respectively. When WP is applied the description of the "A" concepts will only include their labels, thus, the two concepts will be reported as full match by IROM. On the other hand, using WSN extends the description of the both "A" concepts to include "C" and "D" respectively. Therefore, the similarity between the two "A"s will also depends on the similarity between "C" and "D" which are different in this case. As a result, the two "A" concepts will not be recognized as full match.

To conclude, if it is already known that the two ontologies have high syntactical similarities WP is preferable to use. On the other hand, in cases when the compared concepts are not syntactically similar, WSN can extend their descriptions to include new items that can be used to determine whether those concepts are totally different or similar to some degree.

For future work we propose the usage of a measure for calculating the degree of the syntactic overlap between the two ontologies to support the decision of which similarity measure to apply.

Different IR Models: VSM vs. LSA To compare the performance of different IR models, IROM was configured to use WSN and to apply LSA retrieval. We

[4] Average f-measure is defined as the harmonic mean of average recall and precision values of all matched ontology pairs: $avg(f-measure) = \frac{2.avg(recall).avg(precision)}{avg(recall)+avg(precision)}$.

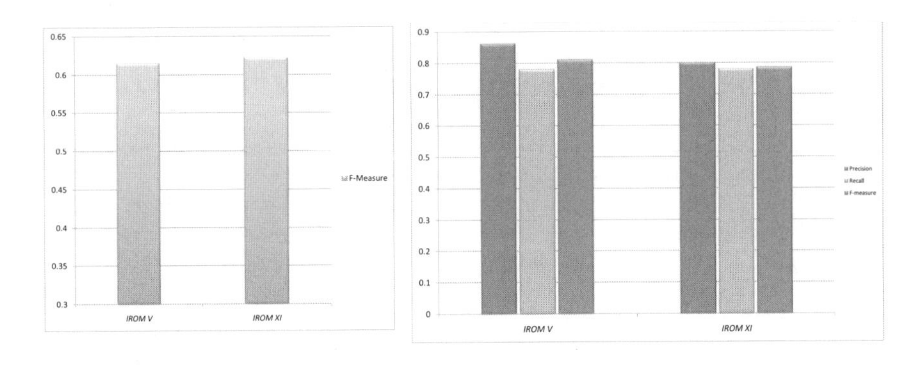

(a) Benchmark Group (248-266) **(b)** Benchmark Group 3xx

Fig. 7. IROM V vs. IROM XI. In (7a) the comparison is based on average f-measure while in (7b) the 3 standard measures are illustrated

call this configuration IROM XI. Two interesting points is to be noticed about the behaviors of IROM XI. First, for ontologies sharing very few entities, LSA provides better matching results than VSM. For example, for the benchmark dataset, IROM XI was able to discover new correspondences for pairs of the third subgroup of 2xx and so provided a better f-measure than that of IROM V (Figure 7a).

On the other hand, for the group 3xx of the same dataset (Figure 7b) as well as for FOAM ontologies, IROM XI was able (for most of the pairs) to discover the same set of correspondences which was found by applying IROM V, however, with slightly lower precision. These results can be explained according to the dimension reduction step of LSA. The absence of a standard method for discovering a best approximation of the LSA space makes it difficult to choose a reduction factor that fits all matching cases.

5 Conclusion and Outlook

In this paper we presented a new ontology matching approach that is based on IR and ontology similarity. Although IR has been used by some matching approaches, its application has always been implicit. IROM provides a deeper insight on the benefit of applying IR models combined with ontology similarity as a weighting system on ontology matching. The evaluation shows that IROM is a promising approach. It is able to align ontologies of different natures with matching performance comparable to that of riper matching approaches. Furthermore, the proposed matching approach provide an infra-structure for a powerful and easily extensible matching system. In future work, we will consider extending IROM to handle the extensional dimension of the matched ontologies and to cover advanced correspondences like 1-n. To improve the precision and the recall additional ontology language features such as comments and property hierarchies (The element `rdfs:subPropertyOf`) will be included to enrich entity descriptions.

References

1. Baeza-Yates, R., Ribeiro-Neto, B.: Modern Information Retrieval, 1st edn. Addison Wesley, Reading (1999)
2. Budanitsky, A., Hirst, G.: Evaluating wordnet-based measures of lexical semantic relatedness. Computational Linguistics 32(1), 13–47 (2006)
3. Ehrig, M.: Ontology alignment: bridging the semantic gap. Springer-Verlag New York Inc., Heidelberg (2007)
4. Euzenat, J.: An API for ontology alignment. LNCS, pp. 698–712 (2004)
5. Euzenat, J., Shvaiko, P.: Ontology Matching. Springer-Verlag New York Inc., New York (2007)
6. Jean-Mary, Y., Shironoshita, E., Kabuka, M.: Ontology matching with semantic verification. Web Semantics: Science, Services and Agents on the World Wide Web 7(3), 235–251 (2009)
7. Jiang, J.J., Conrath, D.W.: Semantic similarity based on corpus statistics and lexical taxonomy. In: International Conference Research on Computational Linguistics (ROCLING X), p. 9008+ (September 1997)
8. Kontostathis, A.: Essential dimensions of latent semantic indexing (lsi). In: Hawaii International Conference on System Sciences, vol. 40, Citeseer (2007)
9. Landauer: Handbook of Latent Semantic Analysis. Lawrence Erlbaum Associates, Mahwah (2007)
10. Letsche, T.A., Berry, M.W.: Large-scale information retrieval with latent semantic indexing (1997)
11. Levenshtein, V.: Binary Codes Capable of Correcting Deletions, Insertions and Reversals. Soviet Physics Doklady 10, 707 (1966)
12. Mao, M., Peng, Y., Spring, M.: An adaptive ontology mapping approach with neural network based constraint satisfaction. Web Semantics: Science, Services and Agents on the World Wide Web (2009)
13. Nagy, M., Vargas-Vera, M., Motta, E.: DSSim–managing uncertainty on the semantic web. In: Proceedings of the International Workshop on Ontology Matching, Citeseer (2007)
14. Resnik, P.: Using information content to evaluate semantic similarity in a taxonomy. In: International Joint Conference on Artificial Intelligence, vol. 14, pp. 448–453 (1995)
15. Seddiqui, M., Aono, M.: Anchor-Flood: Results for OAEI-2009. In: Proceedings of Ontology Matching Workshop of the 8th International Semantic Web Conference, Chantilly, VA, USA (2009)
16. Styltsvig, H.B.: Ontology-based Information Retrieval. PhD thesis, Roskilde University, Denmark (2006)
17. Tang, J., Li, J., Liang, B., Huang, X., Li, Y., Wang, K.: Using Bayesian decision for ontology mapping. Web Semantics: Science, Services and Agents on the World Wide Web 4(4), 243–262 (2006)
18. Wang, P., Xu, B.: LILY: the results for the ontology alignment contest OAEI 2007. In: Proceedings of ISWC 2007 Ontology Matching Workshop, Busan, Korea, Citeseer (2007)
19. Wu, Z., Palmer, M.: Verbs semantics and lexical selection. In: Proceedings of the 32nd Annual Meeting on Association for Computational Linguistics, pp. 133–138. Association for Computational Linguistics, Morristown (1994)

Interoperability for the Design and Construction Industry through Semantic Web Technology

Pieter Pauwels[1], Ronald De Meyer[1], and Jan Van Campenhout[2]

[1] Ghent University - Department of Architecture and Urban Planning,
J. Plateaustraat 22, 9000 Ghent, Belgium
[2] Ghent University - Department of Electronics and Information Systems,
Sint-Pietersnieuwstraat 41, 9000 Ghent, Belgium

Abstract. The domain of architecture, engineering and construction (AEC) has experienced significant improvements with the advent of building information modelling (BIM) applications, which allow AEC specialists to model all information concerning a building design into one three-dimensional building model. Much of these improvements are however generated by the mere availability of such an environment, whereas many more improvements were expected by achieving an appropriate interoperability of information. We are investigating why such an interoperability is not reached fully and consider the semantic web as an alternative approach to reach the targeted interoperability. In this paper, an AEC description framework based on semantic web technology is presented and compared to the BIM approach, after which we indicate how it might solve the issue of interoperability more appropriately. Our evaluation of this investigation indicates the semantic web approach as a valid alternative approach, although considerably more research is needed to show it capable of providing the targeted interoperability of information in the AEC domain.

1 Introduction

The amount and diversity of information is one of the most notable characteristics of a project in the domain of architecture, engineering and construction (AEC). Many domain experts with different backgrounds typically meet within the context of a building project, each of them composing an understanding of the building design and providing a contribution to the project. Each of these experts relies on different software tools for computer-aided design (CAD). Ubiquitous to this context of continuous information flows between experts and their CAD tools, is the point where all these information flows come together. Because this point combines several interpretations of the same subject, it is crucial to maintain a correct understanding and control over these interpretations, so as a sufficiently high level of efficiency can be obtained throughout the design and construction process. Since several years, this point is increasingly addressed by diverse building information modelling (BIM) applications [1]. Within a BIM application, an AEC expert is able to model a three-dimensional BIM model

T. Declerck et al. (Eds.): SAMT 2010, LNCS 6725, pp. 143–158, 2011.

containing all kinds of information about the building designed, including 3D geometry, cost information, material information, etc. In this approach, only one building model is kept at the centre of the AEC design and construction process. Every application should then rely on this central building model so as to achieve interoperability of information throughout the design and construction process within the building design team. By enhancing the communication of information between the software tools deployed by AEC specialists, BIM thus improves the information management process in general.

Recent research [2], however, seems to indicate that BIM merely improves this communication process by supplying the means to construct one three-dimensional information model of the building, and considerably less by enhancing the interoperability of this information between the applications deployed by AEC specialists. BIM is namely mainly used for visualisation, clash detection, building design and the construction of an as-built model and less for communication of information to applications for building analyses and calculations [2]. Such an interoperability is, however, one of the main objectives of the Industry Foundation Classes (IFC), a schema designed and built by the International Alliance for Interoperability (IAI) for the communication of building information between BIM applications [3]. By converting a BIM model to an IFC description in one BIM environment, interoperability should be achieved with any other BIM environment and several of the compatible calculation and simulation applications. However, many problems persist in re-using this IFC building information because it is distorted and/or lost during conversions to and from the IFC schema [4]. This is confirmed by the significant limitations and difficulties we encountered in implementing an energy and acoustic performance checker based on building information in IFC [5]. Although many attribute this problem to a poor implementation practice, we argue in this paper that the root of this poor implementation practice lies in the nature itself of converting semantically rich information. When converting information from one description or representation into another, information distortion or loss almost always occurs, as this is inherently part of the semantic and syntactic difference between these two description schemas.

Because semantic web technology enables the description of information together with its inherent semantics [6], we are currently investigating the adoption of semantic web technology for improving interoperability in the AEC domain as an alternative to the existing BIM plus IFC approach. Instead of relying on a central IFC standard to enhance interoperability, we suggest finding a way to connect diverse alternative schemas together, among which the IFC schema, but also the schema underlying any of the BIM environments, so that one can describe information as he or she wants and still connect it to representations in other schemas. We thus aim at investigating to what extent several alternative descriptions of one and the same building may coexist and be interrelated in one and the same semantic web, presuming that such an interrelated combination of parallel building descriptions might encompass the issues concerning interoperability through IFC.

We start from a general comparison between BIM and IFC technology, and semantic web technology, after which we investigate to what extent semantic web technology may form a valid alternative for the BIM and IFC approach. This investigation is done in the light of our architectural information modelling (AIM) framework, a framework under development for the enhancement of design and building support based on semantic web technology [7]. In this context, we are building a semantic web graph containing all kinds of building information, including geometric information, material information, architectural design intents, etc., after which we investigate how applications may retrieve and meaningfully reuse this information for the various purposes given above. After this discussion of the current AIM framework, we will discuss how parallel building descriptions may be made available using semantic web technology, as an answer to the interoperability issues outlined above.

2 BIM and the Semantic Web

2.1 Limitations in the BIM Approach

The AEC domain involves all kinds of information, covering material characteristics, elementary architectural design notions, legal regulations, three-dimensional parameters, etc. Traditional information systems do not incorporate all this information. At best, an information system incorporates only part of the information needed and provides support based on this information solely. Consequently, numerous applications have emerged focusing on the calculation of building performances, on the checking of the conformance with legal regulations and standards, on the graphic representation and rendering of the design, etc. Because the design and construction process requires the combined input of all this information, AEC specialists are required to deploy a whole range of diverse applications, going from three-dimensional rendering platforms, to CAD applications for stability or energy performance calculations, to internet platforms with disparate resources and additional documentation, etc. Since each of these applications has its own way of describing and managing building information, considerable amounts of information are remodelled and redescribed according to the schema deployed by the targeted information system. The resulting duplicate 3D information models inevitably lead to a significant loss of time and resources and to an increased risk of construction errors and misconceptions in the design [8].

Several approaches have been proposed to tackle this situation, the BIM approach being one of the most notable and successful [1]. Within a BIM platform, a designer is able to model a three-dimensional BIM model containing all kinds of information about the building design, including 3D geometry, cost information, material information, etc. Within the same BIM platform, several applications are available that re-use this information and thereby supply a significant set of calculation and simulation modules. In [2], a high percentage of the respondents confirms the expected benefits of BIM, namely a decrease in building cost and an increase in efficiency, especially for the construction-related phases. Because BIM is, however, mainly used for visualisation, clash detection, building design and the construction of an as-built model [2], these benefits seem not generated by the increased level of interoperability of information, but merely by the availability of an environment to manage building information in one central 3D model as opposed to a collection of 2D CAD drawings. The user remains confined to the schema adopted and thus imposed by the platform developers. It is therefore impossible to describe information that falls out of the original schema, neither is it possible to re-use the information in applications that deploy even a slightly different schema, which both occurs often in the case of the AEC domain.

The development and increasing adoption of the Industry Foundation Classes (IFC) encompasses the latter limitation to a certain extent by providing one neutral schema to describe building information [3]. It was designed and built as a separate EXPRESS schema within STEP by BuildingSMART, formerly known as the International Alliance for Interoperability (IAI) [9]. Its aim is to provide easy communication of construction-related information back and forth between BIM modelling environments and other IFC-compatible software environments. Notwithstanding the significant benefits found in the adoption of IFC [2], most of the expected benefits are not met in real world practice. In several use-cases, information communicated through IFC is found to be distorted or lost, consequently making it far from reusable [4,5]. The conversion process from and to any file format is namely always subject to varying interpretations. In most cases no exact mappings exist between the IFC description of a building and the data schema of the application at hand, making it nearly impossible to achieve the interoperability goal originally targeted. As a result, partners in building projects tend to exchange the IFC model as just another information model, resulting again in multiple building models managed in parallel, containing different information about the same subject.

2.2 Possible Enhancements through a Semantic Web Approach

Because semantic web technology promises the means to connect all kinds of information into one semantic web, including their inherent semantics [6], it might provide for an appropriate alternative approach. It might namely enable the description of information in several distinct, but nonetheless connected graphs, for instance explicitly connecting a building representation in the IFC schema to representations of the same building according to alternative 3D description schemas, such as the X3D [10] or the STL schema [11]. Applications may rely on this web of information as a central information source to provide services as needed for each of the members of a design team. Several existing approaches have already shown how such usage of semantic web technology may also enhance decision support in AEC. The conformance-checking system discussed in [12], for instance, shows how semantic queries may be used to check if a building is conform legal regulations. In [13], an IFCOWL system has been developed to enable the translation of the formal IFC schema into a semantic web graph, thereby enabling an improved partitioning of the information described in IFC [13]. We have extended these initial research efforts with an investigation on the applicability of semantic web technologies in general for building performance checking, more specifically for building acoustics [14], and for the exchange of 3D information within the AEC domain [15]. This research does not focus on the deployment of graphs and ontologies to represent AEC information and improve interoperability, but, instead, elaborates mainly on the usage of rules on top of such graphs for building performance checking and for the conversion of 3D information respectively. The combined merit of having an improved description of the architectural and building information has been discussed in [16] through the presentation of a *"graph-based knowledge specification"* as a basis for a conceptual design system named ConDes. Although the presented system is not based on semantic web technology, it gives a good indication of how architectural and building knowledge may be described semantically and how this knowledge may be reused in advanced reasoning processes for an improved design decision support.

3 A Web of Architectural Information

We created a semantic web of architectural information to simulate how architectural information might be available in a semantic web approach. Several initial test cases illustrate how this web of information might be consumed in the targeted AIM framework, thereby illustrating the applicability of semantic web technology for the AEC domain. This will be documented in the remainder of this paper, thereby leaving out an in-depth discussion of semantic web technologies, as elaborate sources are available elsewhere [17–23].

3.1 AIM Facts

The created semantic web contains information described according to both the IFC ontology and to a newly created AIM schema. Both schemas are to be considered complementary since the IFC ontology focuses on describing a building from the perspective of construction and engineering, while the AIM schema aims at more architectural design concepts, such as style information, history information, design intent information, etc. We started with converting the original IFC schema [3] into an OWL ontology [20], thereby largely following the approach presented in [13]. Using this ontology, an online web service is built and maintained [24] through which file-based IFC models can be uploaded for conversion into IFC/RDF graphs [18]. The converted graphs are made available online through a SPARQL endpoint [23] for query access [25].

As a second part of the AIM graph, several AIM ontologies have been built to enable the structured description of architectural content. These ontologies are constructed solely for investigating the applicability of semantic web technology within the AEC field and can by no means be considered definitive. Central to this set of AIM ontologies is the 'design ontology' (Fig. 1). Using this ontology, one is able to describe design entities and their inherent information. Because a design entity may be linked with very diverse kinds of information, the design

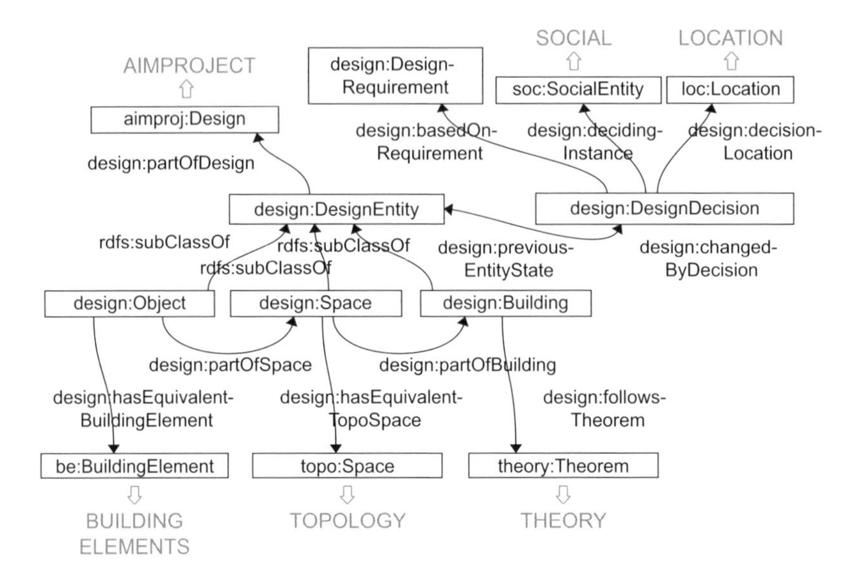

Fig. 1. Graph overview of the design ontology

ontology makes references to several other AIM ontologies focusing on separate knowledge domains in the AEC field concerning building elements, construction types, topology, theory, geometry, people, location, etc.

As shown in Fig. 1, design entities may either be objects, spaces or buildings. An object may be part of a space, and a space may be part of a building. An object may have an equivalent building element, described according to our building element ontology, and a space may have an equivalent topospace, described according to our topology ontology. A design entity may be part of a design and may follow a specific theorem, described using the aimproject and the theory ontology respectively. Using the design ontology, one may also describe design decisions and requirements. When a designer is not satisfied with a specific design entity, for instance because it does not satisfy certain design requirements, a new design entity may be created, following an explicitly described design decision. Both the original and new design entity states are referenced in this design decision instance, thereby explicitly describing the effect of this decision in the RDF graph.

The referenced AIM ontologies allow the description of information stemming from separate knowledge domains. The building element ontology, for instance, enables the description of diverse building elements, making distinctions between doors, beams, walls, etc., and their proper characteristics, such as material-specific characteristics or the way in which they are related to a joined building element. Furthermore, each of these building elements can be linked to a construction type and an IFC representation. This link to an IFC description enables an explicit link between an AIM graph relying on the above AIM ontologies and an IFC graph as it is generated by our IFC-to-RDF converter service [24]. The topology ontology allows the description of spatial topologies, including for instance the description of the visual relation between an adjoining living room and a kitchen. A theory ontology relates design entities with theoretic architectural information, describing for instance which style may be associated to particular elements or which architectural type may be recognized for the overall building.

3.2 AIM Rules

Using the information explicitly available in the AIM graph and the AIM ontologies, extra information may be inferred using plain logic. A simple inference one may want to make is for instance that, `IF` a building floor is accessible from the outside `AND` a room has an entrance door on this floor, `THEN` this room is also accessible from the outside. Using a dedicated semantic rule language, one is able to describe such inference rules containing the condition(s) that need to be met for the conclusion(s) to be true. We previously tested this in the context of acoustic simulations [14], so we will not elaborate on this functionality here. Starting from the IFC ontology and the construction type ontology, we wrote several rule sets in N3Logic that can be used to infer the acoustic performance level of building elements based on the RDF graph describing these elements. Similar to the inference of an acoustic performance level, other inferences may be made as well, as long as the information needed for making these inferences is available in the RDF graph.

3.3 An Example AIM Graph

Using these semantic web technologies, one is able to describe AEC content in a semantic web graph. An example of an RDF graph was built for a design in Antwerp, Belgium. A small part of this graph is given in Fig. 2. It displays part of the description of one of five different designs made during an early design stage, exploring the desired topology in the design.

The graph describes two instances of a `design:Space` concept, how they are influenced by a certain design decision, which design they are part of, and how they are related to other geometry and spaces in the design. Both spaces are related to one instance of a `design:DesignDecision` concept, the one describing the `design:newEntityState` and the other the `design:previousEntityState`. In this case, this describes how, at a specific time during the design process, the `inst:Space_25` instance was replaced by the `inst:Space_42` instance, which is in fact an aggregation of three other spaces, namely `inst:Space_43`, `inst:Space_44` and `inst:Space_45`. This decision is an answer to an instance of the `design:DesignRequirement` concept, i.e. `inst:DesignRequirement_1`. This design requirement describes that each apartment should separate private spaces (e.g. sleeping) and more public spaces (e.g. living room). The result of the design decision is a space which follows the originally described topology of the building, but which is subdivided in an open living space and a secluded sleeping room, separated by a private staircase.

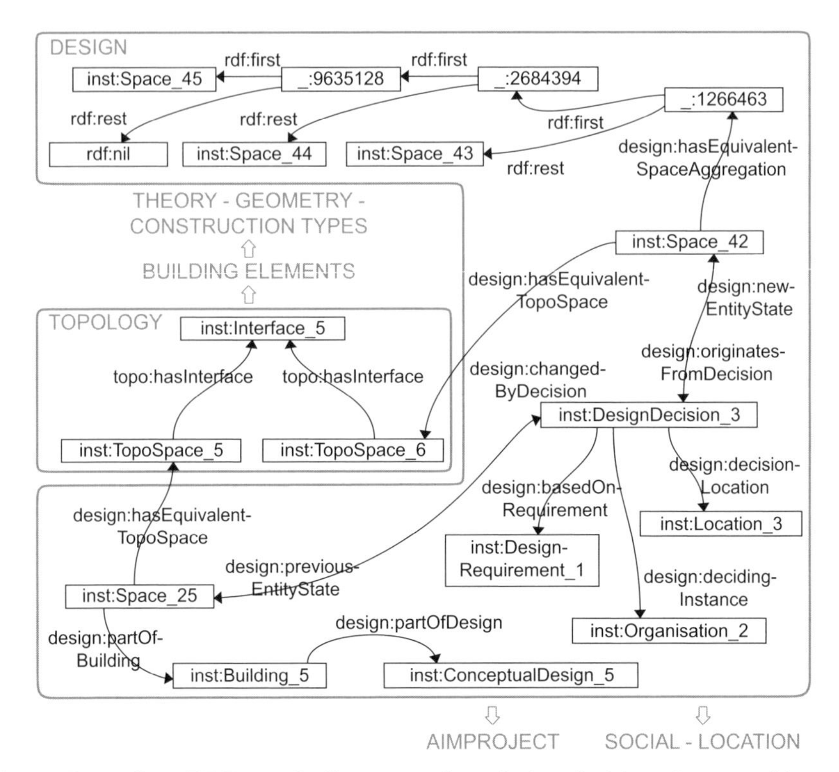

Fig. 2. Part of an RDF graph illustrating how design decisions in an architectural design project may be described

Another part of the same RDF graph, illustrated in Fig. 3, describes a part of the steel structure of the design. This RDF graph is directly related to the RDF graph shown in Fig. 2 through the `inst:AIMProject_1` instance. Every `design:Object` instance (e.g. `inst:Beam_1`) is linked to its equivalent `be:BuildingElement`, thereby explicitly connecting design properties to construction type properties, IFC properties, geometric properties, etc. Figure 3 further illustrates where to include a description of the geometric representation (`ifc: representation`) and the 3D placement (`ifc:objectPlacement`) of a building element, or how one may describe a truss consisting of two girders and two columns. By using a separate instance to describe the connection between two building elements, one is able to describe extra properties linked to this connection. When considering columns and beams, this may for instance introduce the possibility to make an explicit distinction between stiff and elastic joints.

All kinds of other information has been described similar to the way in which it is described above, including material properties, geometric properties, theoretic information, etc. So far, this has resulted in a graph structure of about 100.000 RDF triples. Information that is not considered a direct part of the AEC domain may be connected to this graph as well. This may include for instance geographical information (e.g. GeoNames), people and organisation information (e.g. FOAF) or expert material information (e.g. MATOWL).

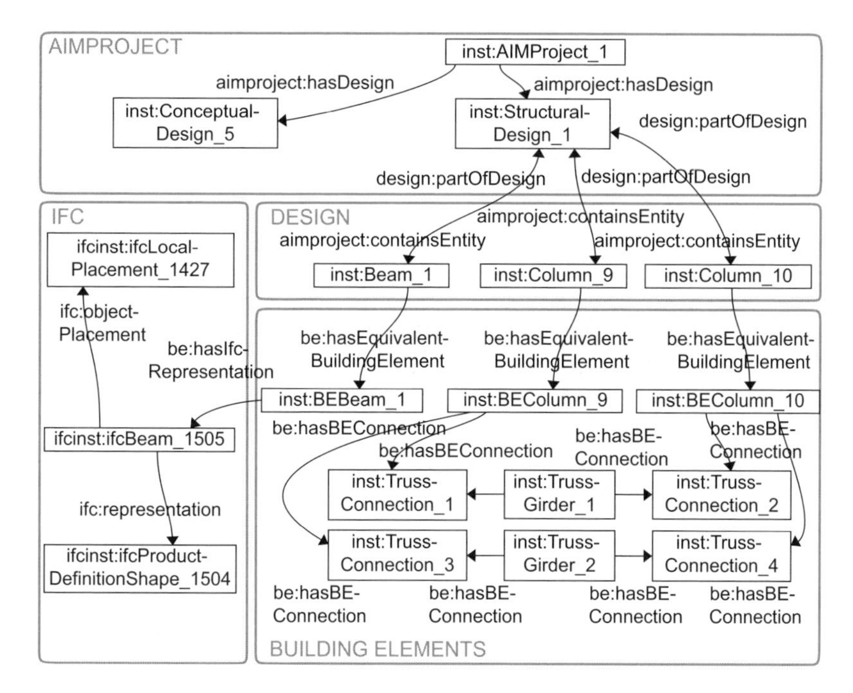

Fig. 3. Part of an RDF graph illustrating how a steel construction may be described as an aggregation of columns and beams

4 Using the Information in Applications

By explicitly describing AEC information in this graph structure, one obtains a semantically rich graph structure following explicitly logical terms. The main difference with existing approaches for design decision support is the ability to re-use this information within logic-based processes, thereby improving query and reasoning possibilities. Because this improves the way in which information can be found and handled, improvements may be found in decision support for the AEC domain as well.

4.1 Advanced Querying of the Information

Using SPARQL [23], one is able to search for very specific information about the architectural design described. The kind of queries processed can be very diverse, depending on the diversity in the semantic structure of the queried RDF graph. The more detailed content is described in the queried RDF graph, the more detailed one can search through this content. For instance, one may search for spaces that were designed as an answer to a given design requirement. Executing such a query on the RDF graph previously discussed (Fig. 2), will then return the `inst:Space_42` instance. Other, more complex queries can easily be constructed using the SPARQL query language. Figure 4 for instance illustrates how one

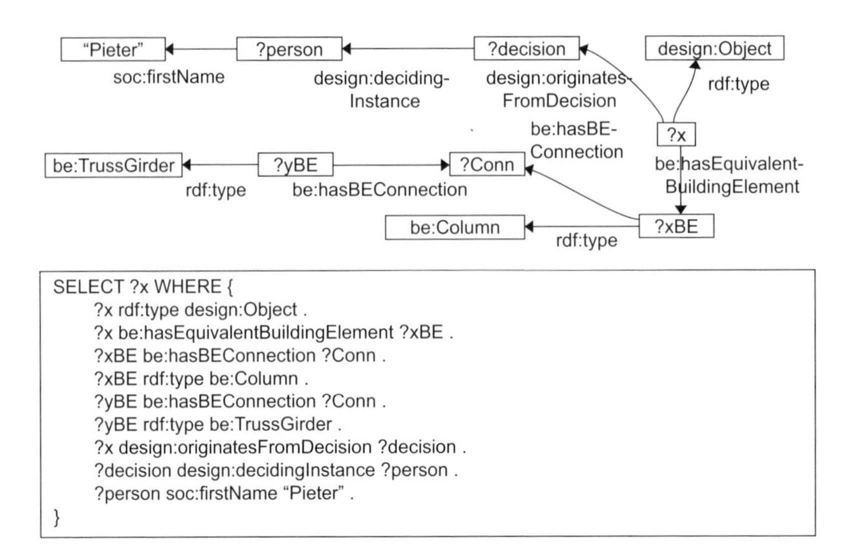

Fig. 4. Query for columns that are connected to a TrussGirder

may search for column designs that originated from a decision made by people with 'Pieter' as their first name, and that have a direct connection with a truss girder element.

4.2 Reasoning about the Information

Using both the explicit (AIM facts) and the implicit (AIM rules) information about a building or a design, one can start a considerable reasoning process. In this case, the derivation of implicit information is done by a reasoning engine and not by the designer. Starting from the building elements ontology, one could for instance infer whether or not Vierendeel beams are present in a design. This is a specific type of beam developed by Arthur Vierendeel previously investigated in its historic context [26]. It is impossible to provide one single definition of a Vierendeel beam, however this kind of beam is most often referred to as a frame(work) with only vertical posts rigidly connected to the top and bottom chord.[1] Starting from this definition of a Vierendeel beam and the information available in the RDF graph of a design, the appropriate rules may be written to infer if this graph contains an instance of such a Vierendeel beam. Figure 5 shows the graph pattern of a rule that may be able to provide this functionality, based on the information in the RDF graph discussed above. It says that `IF` four elements are found, of which two are columns and two are beams, that are connected in a four-sided pattern through stiff joints, `THEN` these elements are part of a Vierendeel beam.

[1] In its original patent, the Vierendeel beam was described by Arthur Vierendeel as a series of rectangular frames *"in which the diagonals are removed and the vertical members rigidly connected to the booms by rounded pieces in such manner that the booms and vertical members form practically one piece"* [27].

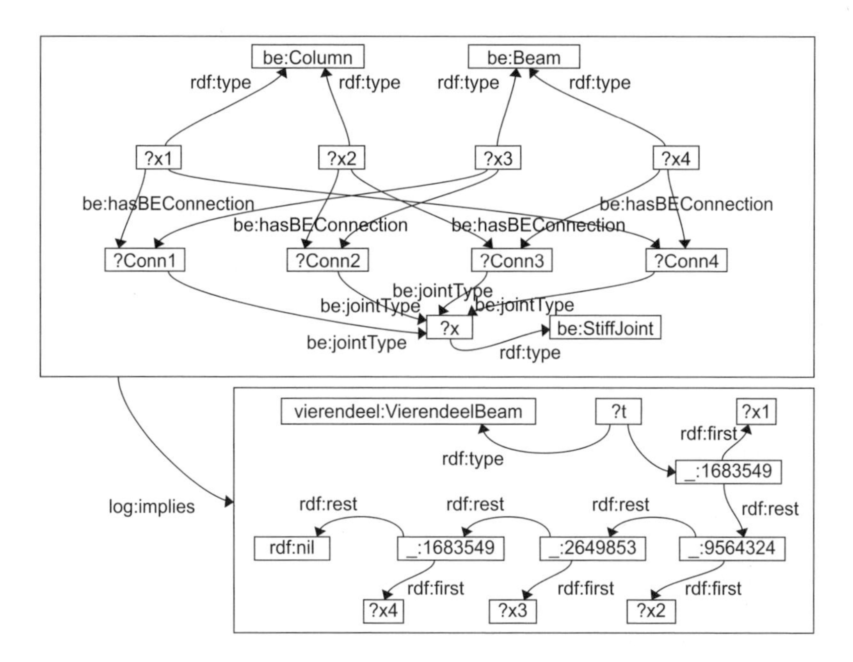

Fig. 5. The above rule describes that this certain configuration of beams and columns constitutes a Vierendeel beam. Note that this graph has no intention to capture the exact definition of a Vierendeel beam. It merely aims at illustrating how extra information may be inferred from an RDF graph within an architectural design process using rules.

When this rule was checked on the RDF graph discussed above using our reasoning engine, several Vierendeel beams were identified, which is of no surprise, considering the rectilinear structure of the building and the way in which a Vierendeel beam was defined (Fig. 5). Obviously, other information may be inferred from the RDF graph as well, based on the information contained in this RDF graph. In a similar way, one could for instance also infer if a room can be reached from within another room, if an architectural theory may be associated to a design, if the design contains significant similarities with other designs (provided that an RDF graph of this second design is available of course), etc. A more detailed overview of how semantic rule languages may be deployed in an AEC context, can be found in [14] and [15].

5 Parallel Descriptions of Information

The above investigation gives an indication of how semantic web technology may provide support in the design and construction process for the AEC domain similar to how BIM provides this support currently. Nevertheless, the most important improvement of deploying semantic web technology over current approaches might be the possibility to reach a more appropriate level of interoperability. In

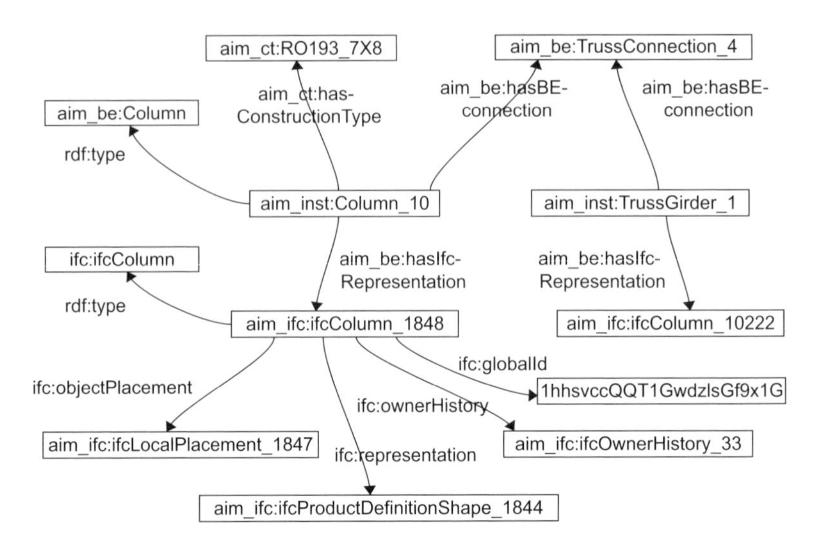

Fig. 6. Part of an RDF graph illustrating how the same column can be described through several distinct subgraphs, in this case as part of an AIM building element graph (`aim_inst:Column_10`) and as part of an IFC graph (`aim_ifc:IfcColumn_1848`).

the light of the discussion about BIM and interoperability earlier, we investigated how we could describe the same concepts and objects in the web of AIM information according to different schemas and interlink these descriptions so that this information may become appropriately reusable or interoperable. An example of parallel descriptions of the same elements can be found in the description of a building element using the AIM building element ontology and the IFC ontology (Fig. 6).

The same concept, namely one of the columns in the design of the Statiestraat, is described through multiple concepts, namely `aim_inst:Column_10` and `aim_ifc:IfcColumn_1848`. Both concepts are linked, with the former further being described using the AIM building elements ontology, and the latter using the IFC ontology. Although these parallel descriptions do not incorporate the same information, they illustrate the interoperability problem appropriately. Whereas IFC provides one standard schema that needs to be reinterpreted and remodelled into another schema, the semantic web allows to combine schemas independently, thereby describing the same concepts according to different schemas. An application on top of this web of information may then rely on information stemming from both schemas as required. For the example in Fig. 6, a BIM environment may rely solely on the information represented by the IFC graph, whereas a simulation environment may rely on very specific parts of the IFC information (e.g. geometry) combined with construction type information following the building element ontology. Changes made in the BIM environment are reflected automatically in the simulation environment, and vice versa, resulting in the required level of interoperability.

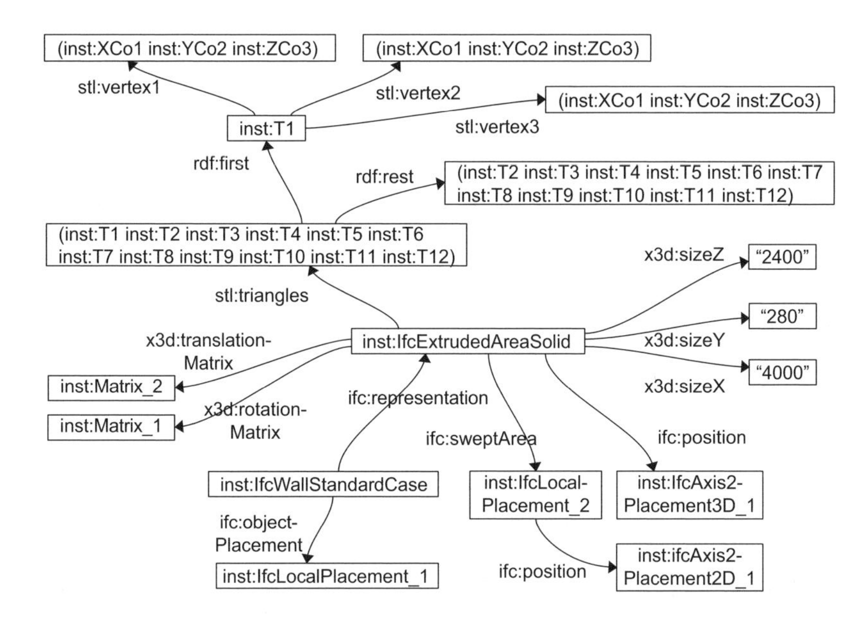

Fig. 7. Part of an RDF graph illustrating how the geometry of a box-shaped wall is described according to the IFC schema (extruded area solid), the X3D schema (rotation and translation matrix) and the STL schema (mesh).

However, in many cases, applications deploy different schemas to represent the same information. This is most often illustrated in the context of 3D geometry. The same geometry can be described in many different ways corresponding to the context in which it is used. One schema may describe a sphere for instance through its centre and radius, whereas another schema may describe it through a circular arc and a central axis, or maybe through a triangular mesh. As is shown in Fig. 7, this information can be combined into one semantic web. Changing part of this information, however, may lead to inconsistencies in the description.

An alternative approach is to rely on rules and an inference engine to infer the duplicate information on-demand, as we indicated earlier in [15]. One geometric description is available in an RDF graph and descriptions following a different schema are generated on demand by a rule engine and a set of inference rules. The information in the alternative schema is thus available implicitly in the RDF graph.

One may for instance consider how an application stores geometry in an RDF graph following the IFC ontology. When combined with the appropriate rule set, namely one that enables the inference of the same geometric information following the STL ontology, one automatically has also this second description at his or her disposal. An application requiring the input of geometry in the STL schema can rely on this inferred STL/RDF graph. Changes made to the geometry in this STL-based application can be stored in a separate RDF graph. A second set of IFC-to-STL conversion rules provides interoperability of information back to the original IFC-based application. This approach could similarly be applied to other, non-geometric information. The initial investigation documented in [15] gives an appropriate overview of this rule-based conversion approach. Further investigations of this approach are however needed to conclude for the practical applicability of this approach.

6 Conclusion

The automation of design and construction in the AEC domain has gone through significant developments and improvements over the past years. The combination of BIM as a central building information management platform and IFC as an interoperability language was in this context suggested as a valid approach for improving efficiency and decreasing the amount of errors and/or misconceptions in the design and construction process. Recent use-case researches indeed indicated such an improvement of the design and construction process. However, it was also shown that these improvements are mainly generated by the mere availability of a central information modelling environment, such as BIM, instead of by the level of interoperability generated by IFC.

As semantic web technology promises a similar level of interoperability for information on the World Wide Web, we started an investigation on the applicability of semantic web technology as a possible alternative approach for inducing improvements through an appropriate level of interoperability. A preliminary architectural information modelling framework was set up, relying on a combination of interconnected facts and rules. We have shown how a design decision support may be set up on top of these facts and rules, similar to the way in which support applications are provided on top of BIM models and/or IFC models.

Additionally, we investigated to what extent the interoperability issues currently present in IFC may be encompassed using such a semantic web approach. This investigation showed us able to describe information using separate description schemas and connect them, thereby enabling a certain level of interoperability. As separate schemas can be combined, one could for instance bring together a schema typically deployed in a BIM environment (e.g. IFC), and a schema typically deployed in a simulation environment (e.g. energy analysis). Finally, we showed how rule languages may enable the combination of schemas that describe the same kind of information differently. The initial test cases in this regard proved successful, although future research is needed to conclude on the practical applicability of the approach.

Acknowledgements. This research is a part of research at the UGent SmartLab research groups, supported by both the Department of Electronics and Information Systems and the Department of Architecture and Urban Planning of Ghent University. The authors gratefully acknowledge the funding support from the Research Foundation - Flanders (FWO).

References

1. Eastman, C.M., Teicholz, P., Sacks, R., Liston, K.: BIM Handbook: A Guide to Building Information Modeling for Owners, Managers, Architects, Engineers, Contractors, and Fabricators. John Wiley and Sons, Hoboken (2008)
2. Becerik-Gerber, B., Rice, S.: The perceived value of building information modeling in the US building industry. ITcon 15, 185–201 (2010)
3. Liebich, T., Adachi, Y., Forester, J., Hyvarinen, J., Karstila, K., Reed, K., Richter, S., Wix, J.: Industry Foundation Classes IFC2x Edition 3 Technical Corrigendum 1, http://www.iai-tech.org/ifc/IFC2x3/TC1/html/index.htm
4. Pazlar, T., Turk, Z.: Interoperability in practice: geometric data exchange using the IFC standard. ITcon 13, 362–380 (2008)
5. Verstraeten, R., Pauwels, P., De Meyer, R., Meeus, W., Van Campenhout, J., Lateur, G.: IFC-based calculation of the Flemish energy performance standard. In: Proceedings of the 7th European Conference on Product and Process Modelling, pp. 437–443. Taylor & Francis Group, London (2008)
6. Berners-Lee, T., Hendler, J., Lassila, O.: The Semantic Web. Scientific American 284, 35–43 (2001)
7. Pauwels, P., Verstraeten, R., De Meyer, R., Van Campenhout, J.: Semantics-based design: can ontologies help in a preliminary design phase? Design Principles and Practices: An International Journal 3, 263–276 (2009)
8. Gallagher, M.P., O'Connor, A.C., Dettbar, J.L., Gilday, L.T.: Cost Analysis of Inadequate Interoperability in the U.S. Capital Facilities Industry. NIST Report GCR 04-867 (2004)
9. International Alliance for Interoperability, http://www.iai-tech.org/
10. Web3D Consortium, X3D International Specification Standards, http://www.web3d.org/x3d/specifications/x3d/
11. 3D Systems, Stereolithography interface specification (1989)
12. Yurchyshyna, A., Zarli, A.: An ontology-based approach for formalisation and semantic organisation of conformance requirements in construction. Automation in Construction 18, 1084–1098 (2009)
13. Beetz, J., van Leeuwen, J., de Vries, B.: IfcOWL: A case of transforming EXPRESS schemas into ontologies. Artificial Intelligence for Engineering Design, Analysis and Manufacturing 23, 89–101 (2009)
14. Pauwels, P., Van Deursen, D., Verstraeten, R., De Roo, J., De Meyer, R., Van de Walle, R., Van Campenhout, J.: A semantic rule checking environment for building performance checking. Automation in Construction 20, 506–518 (2011), http://dx.doi.org/10.1016/j.autcon.2010.11.017
15. Pauwels, P., Van Deursen, D., De Roo, J., Van Ackere, T., De Meyer, R., Van de Walle, R., Van Campenhout, J.: 3D information exchange over the semantic web for the domain of architecture, engineering and construction. Artificial Intelligence for Engineering Design, Analysis and Manufacturing 25 (in press, 2011)
16. Kraft, B., Nagl, M.: Visual knowledge specification for conceptual design: definition and tool support. Advanced Engineering Informatics 21, 67–83 (2007)
17. W3C Semantic Web Portal, http://www.w3.org/2001/sw/
18. Manola, F., Miller, E.: RDF Primer - W3C Recommendation, February 10 (2004), http://www.w3.org/TR/rdf-primer/
19. Brickley, D., Guha, R.V.: RDF Vocabulary Description Language 1.0: RDF Schema - W3C Recommendation, February 10 (2004), http://www.w3.org/TR/rdf-schema/

20. McGuinness, D.L., van Harmelen, F.: OWL Web Ontology Language Overview - W3C Recommendation, February 10 (2004), http://www.w3.org/TR/owl-features/
21. Grant, J., Beckett, D.: RDF Test Cases - W3C Recommendation, February 10 (2004), http://www.w3.org/TR/rdf-testcases/
22. Beckett, D., Berners-Lee, T.: Turtle - Terse RDF Triple Language - W3C Team Submission, January 14 (2008), http://www.w3.org/TeamSubmission/turtle/
23. Prud'hommeaux, E., Seaborne, A.: SPARQL Query Language for RDF - W3C Recommendation, January 15 (2008), http://www.w3.org/TR/rdf-sparql-query/
24. IFC-to-RDF Service, http://ninsuna.elis.ugent.be/IfcRDFService/
25. SPARQL endpoint for IFC/RDF graphs, http://ninsuna.elis.ugent.be/SPARQLEndpoint/
26. Verswijver, K., De Meyer, R.: Past and present characteristics of Vierendeel's poutre arcades. In: Proceedings of the First International Conference on Structures and Architecture (2010)
27. Vierendeel, A.: Girder or Beam for Bridges. USA Patent #639,320, p. 3 (1899)

Dynamic World Modelling by Dichotomic Information Sets and Graphical Inference

With Focus on 3D Facial Pose Tracking

Markus Steffens, Werner Krybus, and Christine Kohring

University of Applied Sciences South Westphalia, Germany

Abstract. This report establishes a novel concept for tracking complex and articulated objects in the presence of high observation uncertainties utilising Markov random fields Markov chains (MRFMCs) and a novel paradigm of modelling visual perception. The approach is rooted in ideas from information fusion and cognitive sciences. The problem is to track non-rigid and articulated objects in the 3D space. The aim is to precisely estimate landmarks with high certainty for fitting accurate object models and secondary states like the orientation under partial occlusions. The targeted system is characterised by a high degree of generality. Previous solutions are relatively limited in robustness and accuracy. The new concept is motivated by the fact that all previous tracking approaches rely on semantic information, that is classified signal signatures, while neglecting all further non-classifiable and thus semantically unrelated information present in the scene herein abstracted as structure. By observing salient cues in structure and by learning and incorporating topological relations between salient cues and semantic features it is intended to tackle the major problem of visual tracking, namely accurate and robust inference in the presence of high observation uncertainties. The notion of the dichotomy of semantic and structure is not covered in previous literature. The new concept constitutes a novel direction in the design and implementation of visual perception and tracking networks. While the ideas of dynamic world modelling and intelligent forgetting stem from principles of information fusion, the principle of fusing semantical with structural information from intelligent exploring is an entirely original contribution and is inspired by ideas from cognitive sciences and linguistics. It is deduced from the inherent yet unrevealed principle of appearance modelling, which is based on incorporating object-related appearance information without classification. In this report the presented system is applied to high-level facial pose tracking and compared to a state-of-the-art reference method.

1 Problem Description and Previous Work

Scene description and analysis in the current context is the process of modelling objects under motion observed in a scene. This is based on the recognition and localisation of a-priori known signal signatures representing pictorial and iconic

T. Declerck et al. (Eds.): SAMT 2010, LNCS 6725, pp. 159–172, 2011.

image features. These can be low-level features such as areal intensity patches or mid-level features like shapes encoded in certain feature spaces. The information extraction and conditioning process follows the traditional sensor fusion framework, that is register, transform, and fuse. It specifically depends on the domain of the application. The next step is to fit an object model onto the extracted features from which further knowledge about an object is to be inferred like position or orientation. A further step is to predict the state of the object for the next time instance incorporating knowledge over the temporal behaviour. Here recognition is meant as a coarse seek process and localisation is a refinement process.

In the context of information fusion there are concepts of dynamic world modeling, focus of attention, and intelligent forgetting. Seminal works are [15,9,37,36].

Another novel direction in spatio-temporal scene analysis is the concept of tracking with structure and so-called spatio-temporal graphs originally coined in the work [23].

In the context of information fusion, probabilistic networks are gaining attention in visual perception and tracking. There the concept of graphical models, as introduced and derived in section (5), is propagated preferably able to handle occlusions though in the appearance domain only. Seminal works are [41,35,5,43,4,13].

The concept of segmentation by partial rigidity and non-rigidity is inspired by the research work in [27,28,29,51]. A comparable concept is non-rigid shape and structure from motion [10,11,45,48,47,44,46] or projective factorisation [24]. In [34] random graphs are used to learn 3D models of unknown objects in a spatio-temporal sense.

In the context of tracking and specifically 3D feature-based tracking, a framework for tracking multiple objects was proposed in [42]. There the aspects of temporal and spatial (stereo) correspondence is discussed, as well as clustering of moving objects based on Kalman motion filters. Other works are closely related to these aspects, cf. [53,49,7]. The contributions covered in this report were developed independent of the referenced later works.

2 Visual Perception and Object Modelling

Methods for scene analysis are often based on exemplar features. Those approaches extract low-level features and compare found candidates with previously learned exemplars, for example patches of eyes encoded in certain feature spaces. These features are used for fitting an object model.

There are generally two types of models. The characteristic of generic models is that the fitting process to the measurements is biased more rigorous to internal generic constraints compared to specific models, since there the constraints are adapted offline or learned to specific exemplars. That is, generic models are unable to cope with larger variations of the objects, or more concrete, the fitting process leads to significantly higher residuals. They are unable to adapt to the observed objects accurately. Contrary, specialised models are only accurate applied to objects within the exemplar space.

Focused applications are characterised by large variations in appearance of articulated objects, for example head and face tracking of drivers. Additionally, the visual information in those unconstrained environments cover higher observation noise such as distractions from out of plane rotations and bad lighting conditions, motion blur, or occlusions. Therefore, current approaches of object modelling will generally fail in these environments.

When analysing the process of fitting an object model to extracted indicator features, it becomes apparent that all indicator features are inherently linked to an a-priori known signal signature or semantic, being an assumed characteristic of the object. For example, geometries explicitly related to eyes or nostrils in facial images. The recognition and localisation step is always restricted to those indicator features. These indicator features are a-priori defined according to certain domain knowledge. Knowledge from other sources, which are not semantically related to the objects under observation, are neglected since they are assumed to contribute no further information. Indeed, those signatures are assumed as noise making the registration step less robust. Further, the scene evolution in a spatio-temporal sense being an inherent characteristic of the objects, or more concrete of the object models, was not incorporated. This aspect was shifted into state prediction, a subsequent process which probabilistically predicts the model states for the next point in time. The temporal evolution of the low-level feature signatures were modelled in so-called appearance models. Here though the target lies on the geometric description of objects.

It can be concluded that the information set incorporated by semantic-based approaches is limited in size due to their inherent concept and it shall be assumed that the set is based on the most accurate information available. Thus the accuracy cannot be improved since any extension of the set means one step from a generic model in direction to a more specific model. The key issue is how to combine and integrate, that is fuse, information from different so-far unrelated sources so that the objects can be modelled more accurately. Relating the semantic object to mid- and high-level contextual information, such as the correlation between location and object or between activity and object, it will be possible to improve the inference process in the presence of insufficient and inaccurate information. The aim of this new approach is to increase accuracy and robustness in the presence of high measurement noise and unstable configurations, such as occlusions. Therefore structure defines any source of information which is not considered by semantically defined object models.

3 Linear Gaussian Error Models

In this report we define an error model consisting of a Gaussian distribution, either in moment form $x \sim p = \mathcal{N}(\bar{x}, P)$ or in information form $p = \widetilde{\mathcal{N}}(h, J)$ where $h = J\bar{x}$ and $J = P^{-1}$. For brevity, we write $\bar{x}$ meaning the mean value, while $\bar{x}$ means the random variable itself. In the following x describes some state, e.g. the 3D location of a point-like object itself consisting of three components.

In this report it is assumed that any entity can be described by an uni-modal, uni- or multi-variate Gaussian distribution, either parameterised by a mean vector x and a co-variance matrix P or a potential vector h and an information matrix J.

Choosing linear Gaussian error models has several reasons. First of all, the parametric description, either in moment form or in information form, is an appropriate approximation for high-level entities having a uni-modal and relatively narrowly peaked (observation) density. Those distributions inherently arise from stereo reconstruction when assuming the data association, that is the correspondence problem, being solved. Further it is convenient for fast and distributed computations. Additionally it can be assumed that these models are more robust compared to multi-modal approximations like particle sets [32,19,17,12] when there is no evidence that the underlying distribution is truly multi-modal.

The derivation of these distributions, that is the propagation of uncertainties yielding the error models, cannot be covered in this report in full length. It should just be mentioned that by the law of linear error propagation the covariance is composed as [8,1]:

$$P = (\prod D_i) P_0 (\prod D_i)^\top \tag{1}$$

where here any D_i is a Jacobian matrix and P_0 is the covariance matrix of some known underlying process. In this report we incorporate implicitly the error models for the sensor-detector chain, facial plane, and the virtual focus of attention. For details please refer to our work in [38].

4 Local 3D-Tracking

Assuming high-level point-like objects to be tracked being part of an augmented object, it is reasonable to assume their distribution of location being Gaussian. The underlying idea is to uncouple the appearance domain, that is, the measurement origin, from the problem of spatial tracking, that is, the target motion uncertainty in location estimation.

The distribution of 3D location of a the i-th point-like at time t object is a Gaussian with mean vector $x_t^{(i)}$ of size 3×1 and a covariance matrix $P_t^{(i)}$ of size 3×3. This distribution is the result of the estimation process, which is in this context motion prediction $p(x_t^{(i)}|x_{t-1}^{(i)})$ and high-level observation $p(z_t^{(i)}|x_t^{(i)})$. Both linear processes are assumed being of Gaussian nature. We adopt the common notation, for details refer e.g. to [32].

The targeted application of facial pose tracking requires the motion prediction by a high-order manoeuvring model. Experiments have shown that constant acceleration models [33] are adequate, especially when combined in tracking networks, cf. section (5). However, fusing velocity and acceleration parameters in a graphical model requires learning of their topological relations. Therefore, here an approximation is made using previous only locations. This leads to the following state prediction according to an autoregressive model:

$$x_t^{(i)} = \frac{5}{2} x_{t-1}^{(i)} - 2 x_{t-2}^{(i)} + \frac{1}{2} x_{t-3}^{(i)} \tag{2}$$

We adapt the notation from [6] and write the augmented state as:

$$\mathcal{X}_{t-1}^{(i)} = [x_{t-3}^{(i)} \quad x_{t-2}^{(i)} \quad x_{t-1}^{(i)}]^\top \tag{3}$$

The local state can thus be estimated by a linear Kalman filter:

$$\mathcal{X}_t^{(i)} = A\mathcal{X}_t^{(i)} + Bw_t^{(i)} \tag{4}$$

$$z_t^{(i)} = C\mathcal{X}_t^{(i)} + Dv_t^{(i)} \tag{5}$$

leading to the following local state distribution:

$$p(x_t^{(i)}|Z_t^{(i)}) = N(x_t^{(i)}, P_t^{(i)}) \tag{6}$$

We choose the matrices B and D as identities, and C maps the most recent location to the measurement $z_t^{(i)}$ where $Z_t^{(i)}$ is the set of all observations up to time t of the local i-th object. The noise components in $w_t^{(i)} \sim \mathcal{N}(0, Q_t^{(i)})$ are learned offline, while the observation (sensor-detector) uncertainties $v_t^{(i)} \sim \mathcal{N}(0, R_t^{(i)})$ are given by the error model of the high-level observation process, here realisations of the process with covariance $R_t^{(i)}$ stemming from stereo reconstruction as discussed in section (3).

5 Spatio-Temporal Gaussian Graphical Models as Tracking Networks

The idea of graphical models [30,21,22,52,18,41,40] is the high-level graph-based abstraction of complex objects by local beliefs of enclosed entities and their topological relations. The local beliefs and the topological relations need to describe the local entities in the same domain, e.g. 3D spatial location and motion parameters.

Here, a local belief at some node describes the probability distribution of a single point-like object, such like the corner of an eye or of a glasses frame or a birth mark. Any topological relation along an edge, called edge potential, describes the spatial distance between two points, that is, the distance of highest probability, the mean, and its uncertainty, the variance.

In Figure (1) a three node model is shown depicting the dichotomic observation origins. According to [26,25] and [20], the probability distribution of a general graphical model is given by a product of singletons and edgewise potentials under the assumption that any entity is conditionally independent on any entity not belonging to its neighbourhood:

$$b(x) \equiv p(x) \propto \prod_{i \in V} \varphi_i(x_i, z_i) \prod_{(i,j) \in E} \psi_{i,j}(x_i, x_j) \tag{7}$$

The singletons over all vertices V are called self-belief or self-potential, and the latter are called edge potential or compatibility potential over all edges

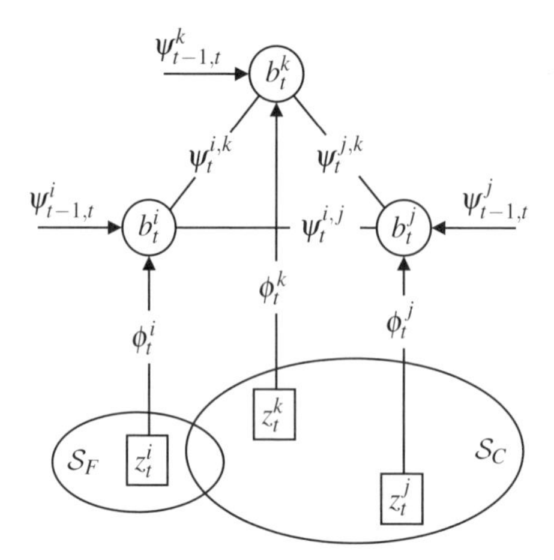

Fig. 1. Concept of dynamic world modelling by dichotomic information sets and graphical inference. The graphical model consists local Markov chains with observations stemming from the set of semanticels $\mathcal{S}_F$ or the set of structurels $\mathcal{S}_C$. The local belief $b_t^i(x_t^i)$ is propagated and inferred from the local posteriors and the topological potentials.

E, where one is interested in decomposing a complex probability distribution $p(x)$ representing the belief $b(x)$ of an object state x into several tractable distributions.

The underlying benefit in Gaussian graphical models is making use of expressing any Gaussian distribution as $\mathcal{N}(x;\bar{x},\Sigma) = \exp\left(Q(x;\bar{x},\Sigma)\right)$ with $Q(x) \propto x^\top\Sigma^{-1}x - 2x^\top\Sigma^{-1}\bar{x} + \bar{x}^\top\Sigma^{-1}\bar{x}$ or $Q(x) \propto -\frac{1}{2}x^\top Jx + h^\top x$ and further taking into account that $\prod_i \exp\left(Q_i(x)\right) = \exp\left(\sum_i Q_i(x)\right)$. The term $x^\top Jx$ shall be called self-information. There are special forms of $Q(x)$ e.g. with $Q(x_i|x_j)$ expressing a conditional density. In these expressions so-called mixed-information $x_i^\top J_{i,j}x_j$ arise with $J_{i,j}$ carrying covariances. For details refer e.g. to [14,31,20,26,25].

By propagating each local node from one time step to the next by a temporal message $M(x_{t-1}^i \rightarrow x_t^i)$, cf. our work in [39], the factors $\varphi_i(x_t^i, z_t^i)$ and $\psi_i(x_{t-1}^i, x_t^i)$ always occur pairwise which can be chosen as the effective posterior probability of x_t^i:

$$\varphi_i(x_t^i, z_t^i)\psi_i(x_{t-1}^i, x_t^i) = p(x_t^i|z_t^i, \{Z_{t-1}^i\}_i) \tag{8}$$

As already mentioned in [40] this is an important property since this allows to perform tracking of the nodes individually and incorporate the effective posterior into the belief propagation scheme as local beliefs $b_{t|t-1}^i(x_t^i)$ becoming updated to $b_{t|t}^i(x_t^i)$ by belief propagation.

This principle can be utilised as a tracking network. The augmentation of Markov Random Fields with Markov Chains (MRFMCs) leads to a tracking network since the local Markov chains, here Kalman filters, are additionally conditioned on all previous histories of related trackers which is emphasised as the local belief $b_{t|t-1}^i(x_t^i) \propto p(x_t^i|z_t^i, \{Z_{t-1}^i\}_i)$.

According to our contributions in [39], we assume having the special case of a Gaussian temporal graphical model, where the state x of the MRFMCs is Gaussian distributed as:

$$x_t \sim \mathcal{N}(\bar{x}_t, P_t) \equiv \widetilde{\mathcal{N}}(h_t, J_t) \tag{9}$$

The diagonal elements of J_t are made up of self-information from the local beliefs as well as of the edge potentials, while the off-diagonal elements are made up of the mixed-information from the edge potentials in case two nodes (i, j) are connected or otherwise zero. The inference, that is the data fusion process over all local beliefs and edge potentials, can be accomplished by loopy belief propagation [41] or for relatively small graphs by brute-force matrix inversion [20]:

$$b_{t|t}(x_t) = \mathcal{N}(J_t^{-1} h_t, J_t^{-1}) \tag{10}$$

The local estimates $x_t^{(i)} \sim \mathcal{N}(\bar{x}_t^{(i)}, P_t^{(i)})$ are obtained as the marginals from $b_{t|t}(x_t)$. The edge potentials $\psi_t^{(i,j)}(x_t^{(i)}, x_t^{(j)})$ are learned as joint distributions $p(x_t^{(i)}, x_t^{(j)} | \Theta_t^{(i,j)})$ conditioned on the history of previous L^2 distance norms in 3D space, the spatio-temporal topology $\Theta_t^{(i,j)}$, between any two connected points i and j utilising variance-weighted regression analysis. For details please refer to our work in [39].

The notion of tracking networks by Markov random fields Markov chains (MRFMCs) is a sound basis for the problem of dynamic world modelling. Their inherent ability is to explore and incorporate new entities and forget lost entities as the scene evolves.

6 Facial Pose Tracking

For robust and accurate facial pose tracking, in this work it is assumed that high-level point-like objects are tracked using a network as described before. Regarding low-level appearance modelling see section (2) for a discussion. The high-level abstraction of a face, that is the object model, is composed of seven semanticels and further structurels, where here the number shall be fixed to four. Thus the tracking network is composed of eleven entities as depicted in Figure (2). The idea is to model the facial pose as the orientation $n \sim \mathcal{N}(\bar{n}, P_n)$ of a facial plane defined by the eye and mouth corners. Fitting of a plane taking the uncertainties of the points into account can be done as described in [50] while for the error propagation please refer to our work in [38].

The locations of the points are assumed being primary states, while the facial plane being composed of the six semanticels eye and mouth corners is assumed being a secondary state. Further the locus of attention $M \sim \mathcal{N}(\bar{M}, P_M)$ in a virtual plane of focus, see Figure (2), is also assumed being an inferred secondary state.

As a reference tracking approach without graphical inference so to compare results an appropriate method is a point-based registration of corresponding sets

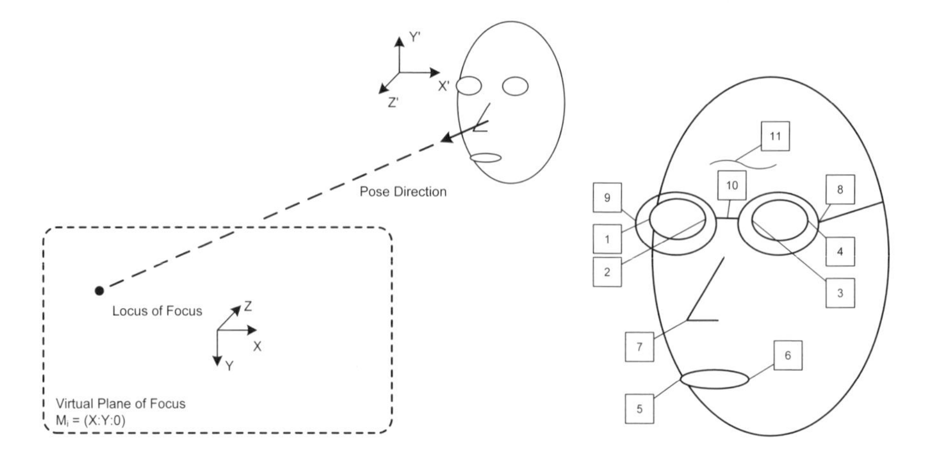

Fig. 2. Left: Diagram of the virtual plane of focus. The facial pose constitutes the line of sight or direction of focus which pierces the virtual plane of focus, here the x-y-plane of the reference camera, which is presumed to cover characteristic positions or areas of interest. Right: Geometric object model showing the semanticels (1-7) and structurels (8-11). The frame has a clockwise orientation where the positive z-direction is headed towards the camera reference frame.

of 3D points in two distinct frames, a world frame and an object-related frame, while parts of the points can be assumed being occluded. This method registers three or more corresponding pairs of points as introduced in [2,3] preferably those with smallest uncertainties. That is, one assumes that these two sets are related by a rigid transform, translation and rotation, possibly superimposed by noise. To allow for non-rigid relations of the facial landmarks, the face model is updated by weighting over the last eight most certain configurations. This is the best heuristic one can choose without further knowledge about the extend of augmentation. The facial pose is proportional to the relative rotation. For the error propagation please refer to our work in [38].

7 Experiments

Sequences of head movements with long phases of partial occlusions as can be expected in front of an instrument panel like a dashboard were analysed; see Figure (3). The average distance between the stereo camera system and the human head was about a half metre. An accurately calibrated stereo camera working with $F = 15$ fps was used for the acquisition of the sequences. The reconstruction accuracy was about 2 milimetres at a distance of 1 metre. The 3D reconstruction was performed on undistorted images, for details refer to [16]. The 2D positions of the landmarks while observable were manually extracted and superimposed by noise with 3 pixels standard deviation for structurels and 5 pixels standard deviation for semanticels contributing to the high-level observation noises $v_t^{(i)}$. For accuracy reasons a circular marker was attached to the

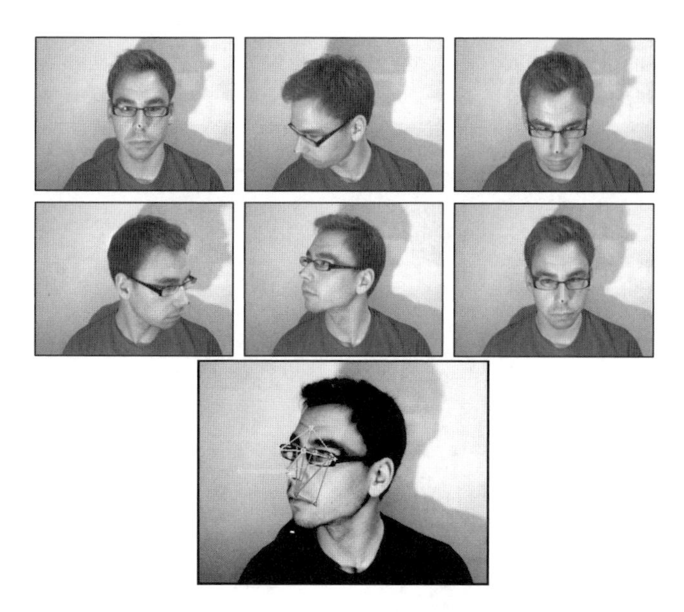

Fig. 3. Left: Exemplary frames from a stereo sequence of 195 frames with 15 fps showing three extreme poses and the inherent occlusions. The sequence describes the poses frontal, down-right, middle-left, up-right, and back to frontal. Right: Landmarks, semantic and structure meshes and pose vector with several entities being occluded.

tip of the nose so to make use of sub-pixel localisation by fitting an corresponding marker model to the image region. The accurate 3D locations of the tip of the nose were used to estimate the motion prediction residuals $w_t^{(i)} = w_t \forall i$ of a constant acceleration model encoded as an autoregressive model.

According to section (4) the motions of local landmarks were predicted by the linear motion model and observed with the mentioned detector noise yielding the local estimates, that is, the effective posteriors $p(x_t^i|z_t^i, \{Z_{t-1}^i\}_i)$. The reference method trivially estimates $p(x_t^i|Z_t^i)$. The topology was learned from L^2 distance norms of sets of previous estimates $\{x_{t-\tau}^{(k)}\}_{\tau=1}^T$ over a window length $T \leq 10F$.

In Figure (4) the traces of attention in the virtual plane of focus is shown along with the propagated uncertainties as recovered by the reference method as well as the proposed system, cf. section (6). Regarding the reference method, when all points are observable, the uncertainties are small compared to those instances where only three or four points are observable. While the local uncertainties of the observable points were relatively small, the propagation of uncertainties into the secondary states lead to relatively high uncertainties in the virtual plane of focus. Contrary, the trace of attention as inferred from the facial plane covers small uncertainties with the proposed tracking network. As can be seen, during turns of the head the uncertainties first increase and become smaller again while more evidence in the edge potentials is again recovered. The same is true when comparing the angles of orientation. While with previous methods only trends

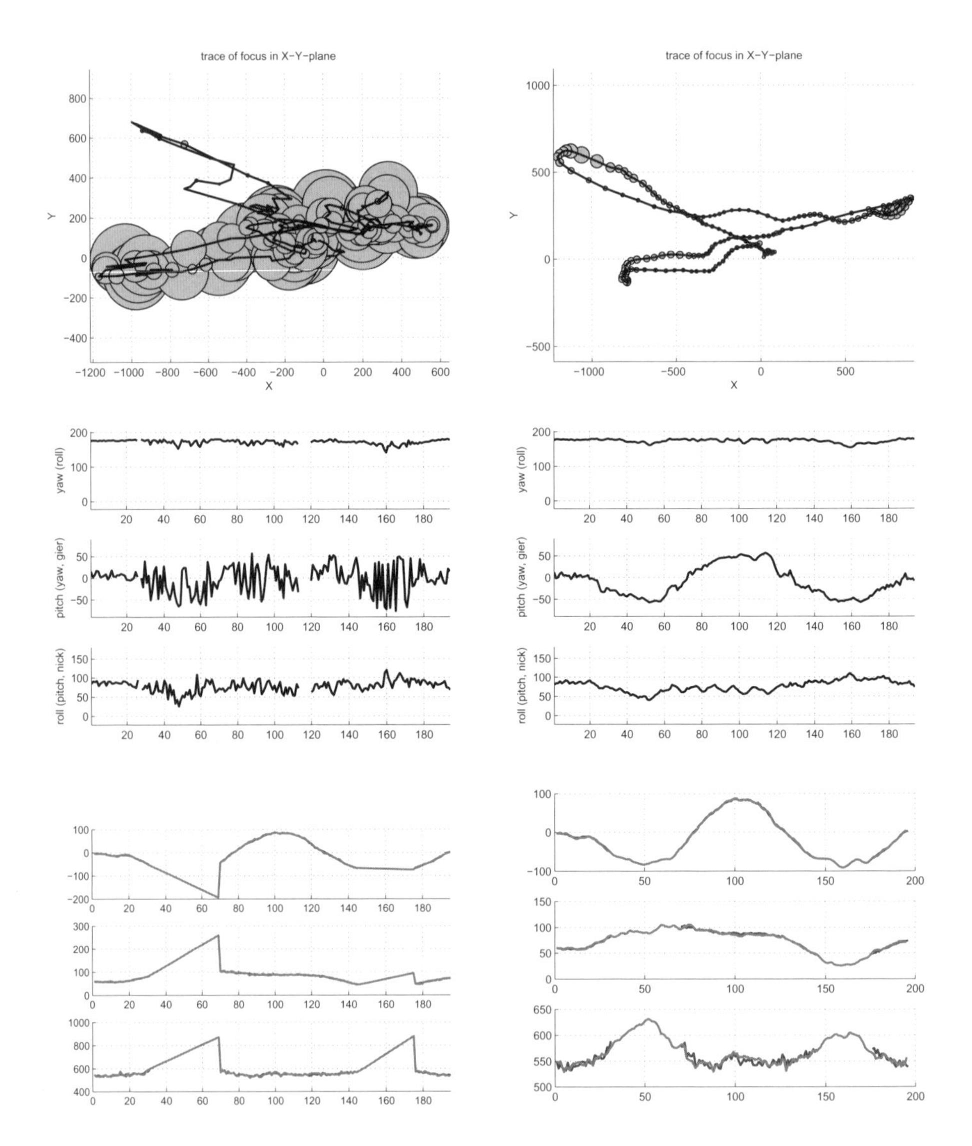

Fig. 4. Left: Reference system. Right: Proposed System with Graphical Model. Top: The trace of attention in the virtual plane of focus plotted in milimetres in the x-y-plane camera reference frame with point-based registration. The green circles indicate the projected absolute uncertainty of each point. The diameter is proportional to the largest semi-axis of the corresponding error ellipsoid of the projected normal vector piercing the plane. The uncertainties were scaled down by a factor of 0.25 (reference system) and scaled up by a factor of 2.5 (proposed system) for visualisation purposes. The trajectory reflects the poses frontal, down-right, middle-left, up-right, and back to frontal again. Middle: Orientation angles in degrees in x-, y- and z-direction. Bottom: Trajectories of fifth semanticel in x-, y- and z-direction in milimetres. The feature became occluded from frame 32 to 69 and 146 to 175.

could be inferred which needed application-specific post-processing, with the novel approach smooth angle estimates can be inferred directly as high-level fusion results taking all available information into account. Looking at the 3D trajectories of a temporarily occluded feature, here a mouth corner, one can see the accuracy by which the proposed system is able to estimate the true trajectory, while the reference system lost track on the local landmark.

8 Summary and Conclusions

The experiments, showing results from graphical inference on semanticels fused with structurels, have clearly shown that the originally formulated approach, that is increasing confidence in classified information by fusion with more accurate unclassified information, directs towards the sought solution of more robust, accurate, general and efficient methods for visual face tracking. The proposed solution makes a big step in direction of solving the introductory stated problems of visual tracking for the application of facial pose tracking, as there are occlusion, scale selection, view dependence, and agile motion. The problem of visual distraction is part of the measurement origin uncertainty which is to be tackled in future research by combining probabilistic appearance models in multi-layer graphical networks, cf. section (9).

This research work further addresses closed analytic solutions to the task of propagation of uncertainties for all involved modules, as there are the sensors, detectors, motion and object models, graphical models, and inferred states. The compactness and simplicity of the linear approximations are relatively accurate, robust and importantly they are able to operate in real-time. This is also true for the proposed Gauss Markov random fields Markov chains. It could be shown that unimodal Gaussian statistics is a legitimate assumption in practical situations. Again it should also be emphasised that these models are able to operate in real-time while being accurate and robust as well as constitute a mathematically profound basis with large potentials for future investigations.

9 Future Work

We propose multi-layer graphical models which learn and estimate low-level appearance entities, that is, taking the low-level measurement origin into account. In the current context the observation process would thus incorporate problems of classification. These processes are naturally multi-modal and require approximations of their distributions such like particle filters. Lower layers propagate association results to high-level layers as presented in this report. In this way we can account for the statistical nature of individual layers more effectively and can expect to derive simpler models, more efficient inference engines [4] as well as to further increase robustness and accuracy.

References

1. Agnew, D., Constable, C.: Geophysical data analysis: Multivariate random variables, correlation and error propagation (2008)
2. Allen, P.K.: 3d photography: Point based rigid registration. Technical report, Columbia Department of Computer Science, Columbia University (2005)
3. Arun, K.S., Huang, T.S., Blostein, S.D.: Least-squares fitting of two 3-d point sets. IEEE Trans. PAMI 9, 698–700 (1987)
4. Bishop, C.M.: A new framework for machine learning. In: Zurada, J.M., Yen, G.G., Wang, J. (eds.) Computational Intelligence: Research Frontiers. LNCS, vol. 5050, pp. 1–24. Springer, Heidelberg (2008)
5. Bishop, C.M.: Pattern Recognition and Machine Learning: Graphical Models. Springer, Heidelberg (2006)
6. Blake, A., Isard, M.: Active Contours. Springer, Heidelberg (1997)
7. Chang, W.-Y., Chen, C.-S., Hung, Y.-P.: Tracking by parts: A bayesian approach with component collaboration. IEEE Transactions on Systems, Man, and Cybernetics 39, 375–388 (2009)
8. Constable, C., Agnew, D.C.: Geophysical data analysis: Statistics (2005)
9. Crowley, J.L., Demazeau, Y.: Principle and techniques for sensor data fusion. Signal Processing 32, 5–27 (1993)
10. Del Bue, A., Agapito, L.: Non-rigid 3d shape recovery using stereo factorization. In: Asian Conference of Computer Vision (ACCV), vol. 1, pp. 25–30 (2004)
11. Del Bue, A., Smeraldi, F., Agapito, L.: Non-rigid structure from motion using ranklet-based tracking and non-linear optimization. IVC 25(3), 297–310 (2007)
12. Doucet, A., Johansen, A.M.: A tutorial on particle filtering and smoothing: Fiteen years later (2009)
13. Du, W., Piater, J.: A probabilistic approach to integrating multiple cues in visual tracking. In: Forsyth, D., Torr, P., Zisserman, A. (eds.) ECCV 2008, Part II. LNCS, vol. 5303, pp. 225–238. Springer, Heidelberg (2008)
14. Gales, M.J.F., Airey, S.S.: Product of gaussians for speech recognition. In: Computer Speech and Language (2006)
15. Hall, D.: Mathematical Techniques in Multisensor Data Fusion. Artech House, Boston (1992)
16. Hartley, R., Zisserman, A.: Multiple View Geometry in Computer Vision. Cambridge University Press, Cambridge (2004)
17. Haug, A.J.: A tutorial on bayesian estimation and tracking techniques applicable to nonlinear and non-gaussian processes. Technical report, The MITRE Corporation (2005)
18. Isard, M.: Pampas: Real-valued graphical models for computer vision. Technical report, Microsoft Research (2003)
19. Isard, M., Blake, A.: Condensation - conditional density propagation for visual tracking (1998)
20. Johnson, J.K.: Estimation of gmrfs by recursive cavity modeling. Technical report, EECS Dept., MIT (2004)
21. Jordan, M.I., Weiss, Y.: The Handbook of Brain Theory and Neural Networks. Graphical models: Probabilistic inference. MIT Press, Cambridge (2002)
22. Jordan, M.I.: An introduction to probabilistic graphical models. Technical report, University of California, Berkeley (2003)
23. Kropatsch, W.: Tracking with structure in computer vision twist-cv. Technical report, Patter Recognition and Image Processing Group, TU Wien (2005)

24. Li, T., Kallem, V., Singaraju, D., Vidal, R.: Projective factorization of multiple rigid-body motions. In: CVPR 2007, pp. 1–6 (2007)
25. Malioutov, D.M.: Approximate Inference in Gaussian Graphical Models. PhD thesis, Dept. of Electrical Engineering and Computer Science, Massachusetts Institute of Technology, Cambridge (2008)
26. Malioutov, D.M., Johnson, J.K., Willsky, A.S.: Walk-sums and belief propagation in gaussian graphical models. Journal of Machine Learning Research 7, 2031–2064 (2006)
27. Mills, S.: Stereo-motion analysis of image sequences. In: Proceedings of Digital Image & Vision Computing: Techniques and Applications (DICTA), pp. 515–520 (1997)
28. Mills, S., Novins, K.: Graph-based object hypothesis. New Zealand Journal of Computing 7, 21–29 (1998)
29. Mills, S., Novins, K.: Motion segmentation in long image sequences. In: Proceedings of the British Machine Vision Conference, pp. 162–171 (2000)
30. Murphy, K.P.: An introduction to graphical models. Technical report, University of British Columbia, Vancouver, Canada (2001)
31. Newman, P., Leonard, J.: A matrix oriented note on joint, marginal, and conditional multivariate gaussian distributions. Technical report, Massachusetts Institute of Technology (2006)
32. Ristic, B., Arulampalam, S., Gordon, N.: Beyond the Kalman Filter: Particle Filters for Tracking Applications. Artech House Publishers, Boston (2004)
33. Rong Li, X., Jilkov, V.P.: Survey of maneuvering target tracking. part i: Dynamic models. IEEE Transactions on Aerospace and Electronic Systems 39, 1333–1364 (2003)
34. Sanfeliu, A., Serratosa, F.: Learning and recognising 3d models represented by multiple views by means of methods based on random graphs. In: Proceedings International Conference on Image Processing, ICIP (2003)
35. Sigal, L., Zhu, Y., Comaniciu, D., Black, M.J.: Tracking complex objects using graphical object models. In: Jähne, B., Mester, R., Barth, E., Scharr, H. (eds.) IWCM 2004. LNCS, vol. 3417, pp. 223–234. Springer, Heidelberg (2007)
36. Sinha, A., Chen, H., Danu, D.G., Kirubarajan, T., Farooq, M.: Estimation and decision fusion: A survey. Neurocomputing 71, 2650–2656 (2008)
37. Smith, D., Singh, S.: Approaches to multisensor data fusion in target tracking: A survey. IEEE Transactions on Knowledge and Data Engineering 18(12), 1696–1710 (2006)
38. Steffens, M., Krybus, W., Kohring, C.: Linear gaussian error models from component matrices for 3d graphical tracking networks. In: Submitted to SAMT 2010 (2010)
39. Steffens, M., Krybus, W., Kohring, C.: Spatio-temporal gaussian graphical models as tracking networks. In: Submitted to SAMT 2010 (2010)
40. Su, C., Huang, L.: Spatio-temporal graphical-model-based multiple facial feature tracking. EURASIP Journal on Applied Signal Processing 13, 2091–2100 (2005)
41. Sudderth, E.B., Ihler, A.T., Freeman, W.T., Willsky, A.S.: Nonparametric belief propagation. In: Conference Proceedings Computer Vision and Pattern Recognition. IEEE, Los Alamitos (2003)
42. Tang, C.-Y., Hung, Y.-P., Shih, S.-W., Chen, Z.: A 3d feature-based tracker for multiple object tracking. In: Proceedings of the National Science Council, Republic of China, Part A: Physical Science and Engineering, pp. 151–168 (1999)

43. Taycher, L., Fisher III, J.W., Darrell, T.: Combining object and feature dynamics in probabilistic tracking. Computer Vision and Image Understanding 108, 243–260 (2007)
44. Vidal, R., Abretske, D.: Nonrigid shape and motion from multiple perspective views. In: Leonardis, A., Bischof, H., Pinz, A. (eds.) ECCV 2006. LNCS, vol. 3952, pp. 205–218. Springer, Heidelberg (2006)
45. Vidal, R., Ma, Y.: A unified algebraic approach to 2-d and 3-d motion segmentation. In: Pajdla, T., Matas, J(G.) (eds.) ECCV 2004. LNCS, vol. 3021, pp. 1–15. Springer, Heidelberg (2004)
46. Vidal, R., Ma, Y., Soatto, S., Sastry, S.: Two-view multibody structure from motion. IJCV 68(1), 7–25 (2006)
47. Vidal, R., Ravichandran, A.: Optical flow estimation and segmentation of multiple moving dynamic textures. In: CVPR 2005, pp. II: 516–521 (2005)
48. Vidal, R., Singaraju, D.: A closed form solution to direct motion segmentation. In: CVPR 2005, pp. II: 510–515 (2005)
49. Wang, P., Ji, Q.: Robust face tracking via collaboration of generic and specific models. IEEE Transactions on Image Processing 17, 1189–1199 (2008)
50. Weingarten, J.W., Gruener, G., Siegwart, R.: Probabilistic plane fitting in 3d and an application to robotic mapping. In: IEEE International Conference on Robotics and Automation (2004)
51. Yang, M., Wu, Y.: Granularity and elasticity adaptation in visual tracking. In: IEEE Conference on Computer Vision and Pattern Recognition, CVPR (2008)
52. Yedidia, J.S., Freeman, W.T., Weiss, Y.: Understanding belief propagation and its generalizations. Technical report. MIT, Cambridge (2001)
53. Yu, T., Wu, Y.: Decentralized multiple target tracking using netted collaborative autonomous trackers. In: IEEE Computer Society Conference on Computer Vision and Pattern Recognition (CVPR), vol. 1, pp. 939–946 (2005)

Use What You Have: Yovisto Video Search Engine Takes a Semantic Turn

Jörg Waitelonis, Nadine Ludwig, and Harald Sack

Hasso-Plattner-Institute Potsdam,
Prof.-Dr.-Helmert-Str. 2-3, 14482 Potsdam, Germany
{joerg.waitelonis,nadine.ludwig,harald.sack}@hpi.uni-potsdam.de
http://www.hpi.uni-potsdam.de/

Abstract. The phenomenal increase of online video content confronts the consuming user with an immeasurable amount of data which can only be accessed with sophisticated multimedia search and management technologies.

Usual video search engines provide a keyword-based search, where lexical ambiguity of natural language often leads to imprecise and incomplete results.

Semantics of keywords and metadata has to be determined to overcome these shortcomings to provide high precision and high recall.

In this work, we show how to gradually transform the video search engine Yovisto from a simple keyword-based search engine to a fully-fledged semantic video search engine simply by using the existing search engine infrastructure based on Lucene augmented by simple semantic metadata.

1 Introduction

There is a continuous increasing demand for online videos in recent years. With more than 14.6 billion video downloads from YouTube[1] in May 2010 only in the USA[2], video and multimedia is becoming the predominant part of internet traffic.

Besides, the entire World Wide Web (WWW) with a growing amount of realtime data has risen beyond any expectations. For the user, the only way to orientate oneself in this huge amount of data and to find the needle in the haystack is making use of search engines.

But, also web search engines are reaching their limits. With the sheer quantity of retrieved documents the user is faced with an almost insolvable task of deciding, whether the desired document really is among the result list or if the result set is complete at all. A simple Google[3] search often returns millions of documents ranked by link popularity. But, what if the desired result is not among the first few result list pages?

[1] http://www.youtube.com/

[2] http://www.comscore.com/Press_Events/Press_Releases/2010/6/comScore_Releases_May_2010_U.S._Online_Video_Rankings

[3] http://www.google.com/

T. Declerck et al. (Eds.): SAMT 2010, LNCS 6725, pp. 173–185, 2011.

Traditional web search engines are based on keywords, i.e. keywords are extracted from web documents for distinctive representation. The user tries to identify the document(s) she is looking for with a query phrase consisting out of one or more keywords, which she supposes to match the keywords being extracted and indexed by the search engine. Unfortunately, besides different user context and pragmatics, natural language suffers from inherent ambiguities that cause deficits in search engine recall and precision.

One way to cope with these ambiguities is to enable a better understanding of the meaning of the web document's content as well as of the user's query string.

Semantic Web technologies intend to make implicit meaning of content explicit by providing suitable metadata annotations based on formal knowledge representations.

Instead of extracting keywords from documents the document content is mapped to distinct entities and classes to avoid ambiguities and to enable higher recall and precision of search results. Content-based search based on Semantic Web technologies often is referred to as 'Semantic Search'. Besides, the problem of different contexts and pragmatics remains, but is not taken into consideration for this paper.

But, do we need an entire new search engine architecture for enabling semantic search? We will show how an existing keyword-based search engine can be augmented by simple means to become a fully-fledged semantic search engine capable of content-based retrieval. To achieve this, we will show how to enable a mapping of index keywords, user provided tags as well as the terms in the user's query string to Semantic Web entities and classes with the help of Linked Open Data, linguistic and lexical resources. Entities and classes in the Semantic Web are identified by URIs (Uniform Resource Identifiers) that will be utilized instead of former text-based keywords within an existing open source enterprise search platform to enable semantic search. The paper will give a step-by-step description on how to turn a keyword-based search engine into a semantic search engine demonstrated at the video search engine Yovisto[4].

The rest of the paper is structured as follows: Section 2 will summarize related work on keyword-based search, multimedia search, and semantic search with a focus on entity mapping and disambiguation. In Section 3, we will give a step-by-step description on how we achieved the turning into a semantic search engine with onboard means. Section 4 concludes the paper with a short summary and an outlook on open problems and future work.

2 Related Work

2.1 Keyword-Based Search

Since Google's conquest of the web search engine universe we know that the quality of keyword-based search mainly depends on the ranking of search results, but also lacks on problems coming with the peculiarities of natural language.

[4] `http://www.yovisto.com`

Synonyms and homonyms cause incomplete and inaccurate search results. Keyword-based search presumes that the user knows the exact keyword for describing the document she is looking for. If the user does not know the appropriate keyword or if she is trying to find the right documents to answer more complex query tasks, the search objective can become unattainable. Query expansion and suggestions help the user to refine the search results and enable a step-by-step convergence [3]. Filtering methods such as facetted search categorize search results and enable to limit the results to subsets for better overview [14]. Beyond the simple keyword lookup, the user can reiterate a sequence of queries while refining and/or extending the query depending on the content of the obtained results. Search scenarios sophisticated and complex like this are referred to as exploratory search [7].

2.2 Multimedia Search

The same inadequacies can be found in the more and more emerging multimedia search. General video search engines such as YouTube support a keyword-based search within the textual metadata provided by the users, accepting all the shortcomings caused by e.g. synonyms and homonyms. For example, a search for "history of golf" will result in documents containing a variety of outcomes for 'Golf' (e.g. sports, car) and by refining the search phrase to "history of golf car" it is astonishing that the top ranked videos are about the 'golf cart', which is obviously not a product of Volkswagen, but has at least four wheels. The ranking of the correct results is thwarted by the automated query completion. However, multimedia search lacks the same problems as keyword-based text search. Compared to textual search, multimedia search allows varied visualizations [15,4].

The video search engine Yovisto is specialized on lecture recordings and scientific presentations and provides a time-dependent video index. With sophisticated video analysis techniques (such as automated scene detection, intelligent character recognition, etc.) in combination with collaborative user annotation Yovisto provides scene accurate access to more than 10.000 videos with pinpoint accuracy [5,12]. To overcome some of the the shortcomings of keyword-based search, Yovisto deploys an exploratory search navigation to allow the user to browse the repository beyond simple fact-finding and page-turning [18,19,20].

2.3 Linked Data and Semantic Search

Yovisto enables the exploratory search by applying *Linked Open Data*[5] (LOD). Linked Data means to expose, share and connect pieces of data, information, and knowledge on the Semantic Web using URIs for identification of resources and RDF[6] as structured data format [1]. The Linked Open Data project aims to publish and connect open but heterogeneous databases by applying the Linked Data principles. One of the most important interlinking LOD hubs is *DBpedia* [2],

[5] http://linkeddata.org/

[6] http://www.w3.org/RDF/

which publishes encyclopedic information from the famous Wikipedia[7]. DBpedia currently provides information about more than 3.4 million "things" with over 1 billion "facts"[8]. Furthermore, *Wortschatz Leipzig* in an automatically compiled thesaurus, which collects large volumes of natural language text from the WWW and applies sophisticated linguistic and statistic analysis in large scale to provide thesaurus information. Wortschatz Leipzig supports more than 17 languages and is also publicly available as RDF Linked Data [13]. The aggregation of all LOD data sets is denoted as *LOD-Cloud*.

To access and to make use of the Semantic Web information, search engines need to index semantic data. Commonly, this can be achieved by storing the data in RDF databases (triple stores). This allows to query the data with structured languages such as SPARQL [11]. Search engines such as *sindice* [8,17] or *swoogle*[9] are crawling the Semantic Web to obtain a data set as large as possible and provide a label and entity based search. Those search engines are already operating on the existing Semantic Web, while this work focusses on how to turn a non-semantic web search engine into a semantic web search engine. Therefore, the already existing keyword-based textual metadata has to be mapped to Semantic Web entities.

The most challenging problem on mapping data to Semantic Web entities is the existence of ambiguous names and thus resulting in a set of entities, which have to be disambiguated. Related work fields are word-sense disambiguation in text documents, named entity (reference) resolution, and feature based entity matching amongst others. The presence of assumed same named entities in different triple stores of the LOD-Cloud necessitates similarity based comparison of the entities and their respective features or properties [16]. In the context of named entity resolution in text documents the semantic information needed for disambiguation of potential entities has to be extracted automatically and compared to adequate knowledge resources [9]. Further research approaches are using the LOD-Cloud itself as RDF graph to find relations between entities co-occurring in a text maintaining the hypothesis that disambiguation of co-occurring elements in a text can be obtained by finding connected elements in an RDF graph [6].

3 How to Turn Yovisto into a Semantic Video Search Engine

Turning Yovisto from a keyword-based search engine to a semantic video search engine needs several steps. Fig. 1 shows an overview of the process sequence to enable semantic search with Yovisto. Yovisto's time-dependent as well as time-independent metadata is mapped to entities of LOD, respectively DBpedia (a). The Wortschatz Leipzig co-occurence network is applied to support entity disambiguation and ranking of entity candidates (b). Within the indexation process the old keyword index is extended by additional semantic information (c). On the users hand, the query is disambiguated with an auto-suggestion select box,

[7] http://www.wikipedia.org/

[8] http://dbpedia.org

[9] http://swoogle.umbc.edu/

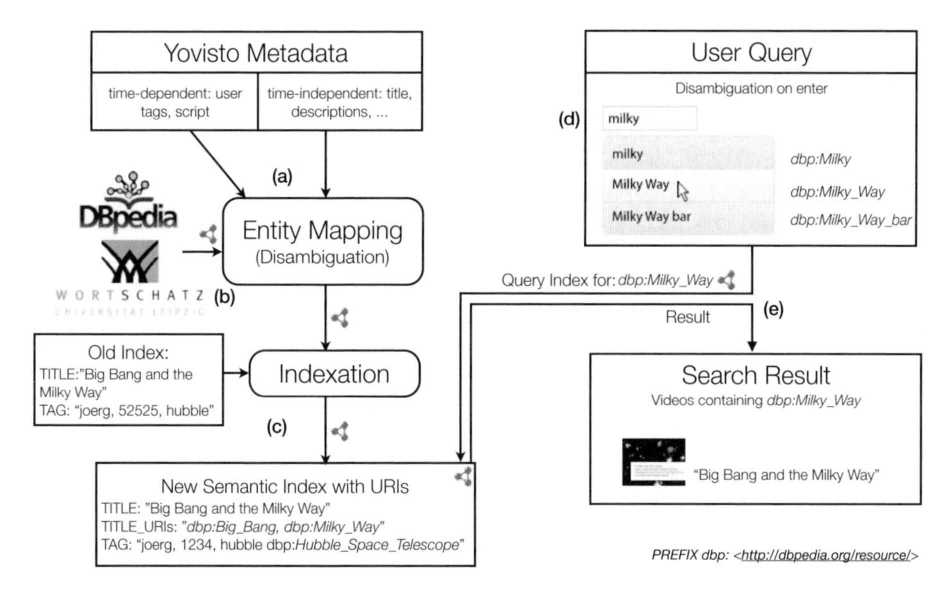

Fig. 1. Overview of the processes of semantic search with Yovisto

providing the URIs of the requested entity (d). Finally, the index lookup returns a list of documents containing the entity (e).

According to the Linked Data principles, Yovisto's metadata has been linked to the LOD-Cloud by connecting organizations, categories, and persons (speaker) to DBpedia entities including a mapping of Yovisto's metadata structure to an appropriate ontology[10] representation [19]. To ensure interoperability, mappings and references to existing ontologies have been used whenever possible. All metadata has been published through a SPARQL-endpoint[11] and embedded via RDFa in Yovisto's web pages.

The mapping of video content to LOD entities is the subject of this paper. The video content is described as well by metadata referring to the entire video (e.g. title) as also by information assigned to a distinct time position within the video. Among these time-dependent metadata are collaboratively generated user tags, user comments, and text extracted from the simple video frames. The following section deals with the entity mapping of both kinds of metadata.

3.1 Entity-Mapping

Currently, user-generated tags and automatically extracted data are provided as text elements essentially for the keyword-based search. To transform the keyword-based search engine to a semantic video search engine the text elements need to be mapped to semantic web entities to set a certain segment respectively a whole video in the context of the LOD-Cloud. The following section describes how to use the existing tags and keywords to create a semantic context mapped

[10] http://www.yovisto.com/ontology/
[11] http://sparql.yovisto.com/

to entities of the LOD-Cloud. An overview of all steps of the entity mapping process are shown in Fig. 2.

Term and Group Mapping. Yovisto provides several methods to equip a video with metadata: video-related keywords and video-time-related keywords and tags. Video-related keywords are supplied by the user who uploaded a video to the portal and this metadata is applied to the whole video. Video-time-related metadata are on the one hand keywords that are automatically extracted from the video on a certain timestamp (e.g. by OCR methods) and on the other hand user-generated tags also on a certain timestamp in the video.

Metadata exist in a certain context and have to be mapped to Semantic Web entities within this context. We define the context of a term (tag or keyword) in a video as the tags and keywords at and around the timestamp in the video the term is occurring at. All tags and keywords are stored separately with timestamp and video ID in the database.

In a preprocessing step we generated a "term mapping table" by using labels, normalized URI-suffix, and labels of redirects and assigning them to the corresponding DBpedia URI. These terms are single words, as well as composed groups of words e.g., names ("Albert Einstein").

The first step to map the metadata of a video to Semantic Web entities is to gather all tags/keywords of a context and build groups to be able to also map groups of tags to entities. Let us consider the following example: a user tagged "hubble", "deep", "space", "exploration", "field" in that order. In that context it might be useful to allocate the DBpedia entities for "deep space exploration", "hubble" (space telescope), and "hubble deep field". Therefore a variation without repetition (to avoid combinations of a tag with itself) on this set of all tags in this context has to be constructed. In this example the set contains $n = 5$ elements and groups of $k = 1, 2, 3$ elements have to be built.

$$(n)_k = \frac{n!}{(n-k)!}$$

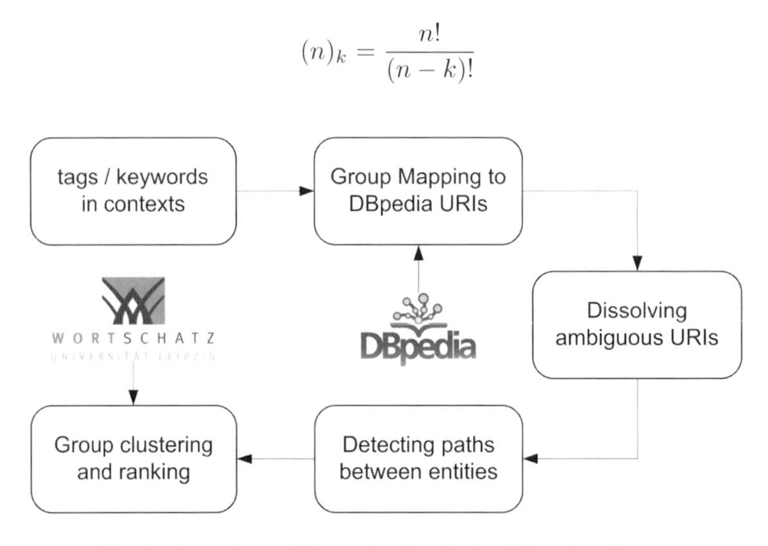

Fig. 2. Overview of Steps in the Entity Mapping Process

For our example, this leads to $(n)_k = 85$ combinations in total. However not all combinations are needed, only the groups that match to a term (from the term mapping table) are stored with their respective URI. This term matching adds up to 13 URIs. As every matching group is stored, there are also URIs for single terms like "field". Such single terms are often assigned to ambiguous URIs and after resolving these ambiguous URIs (also for the terms "deep", "field", "deep space", "hubble", "space") the result set of possible applicable URIs at this timestamp in this video contains 152 URIs. This enormous amount of matching URIs to a set of only five tags in a context leads to the inevitable problem of disambiguation and ranking of the matching URIs.

Problems of ambiguous URIs in DBpedia: In the example mentioned above we only disambiguated the resulting URIs in one "lap". The 152 URIs still contain three ambiguous URIs leading to 24 additional URIs. That means the algorithm for resolving ambiguous URIs has to be run repeatedly taking into account that DBpedia contains "circles" of disambiguations referencing back to already resolved URIs and these circles have to be detected and erased. For performance reasons we constrain the process of resolving ambiguous URIs to two runs.

Currently, Yovisto maintains approximately 22.936 user tags in 9271 contexts (timestamps) in 517 videos. The tags were mapped as single words and groups of 2 and 3 words – 30.417 groups were mapped to DBpedia URIs. The ambiguous URIs were resolved in a first run resulting in 162.523 URIs, after a second run of resolving 192.056 URIs were found.

Disambiguation of URIs. The result set of URIs mapped to the matching groups of tags contains duplicate URIs with different matching groups - e.g., the matching groups "deep field" and "hubble deep field" were both assigned to the URI `<http://dbpedia.org/resource/Hubble_Deep_Field>`. So these duplicate records can be filtered by deleting the records with synonyms as matching groups. In the example, the URI contains as label the term "hubble deep field", so that the record with the matching group "deep field" can be removed from the result set.

For the further disambiguation steps of the diverse URIs in one matching group we assume that entities that are related to each other are also referenced over few properties in the LOD-Cloud (sample relationships of the example entities "Hubble Space Telescope", "Hubble Deep Field", and "Deep Space Exploration" are shown in Fig.3. Therefore, the applicable URIs from the set of URIs for one matching group should be linked to one or more other URIs of the context. We consider an algorithm, which is detecting paths (with properties and other entities as nodes) between the entities of a context. The result set of this path-detecting algorithm is one or more sub-graphs of the LOD-Cloud. Because the amount of properties in the LOD-Cloud is quite large, we need to identify the most meaningful properties that link related entities in a certain context. Also, the degree of relationship - conveyed by the length of the path between two entities - has to figured to meet the right measurement to find both entities from the same context and filter out relations that are too distant to be considered from the same context [6].

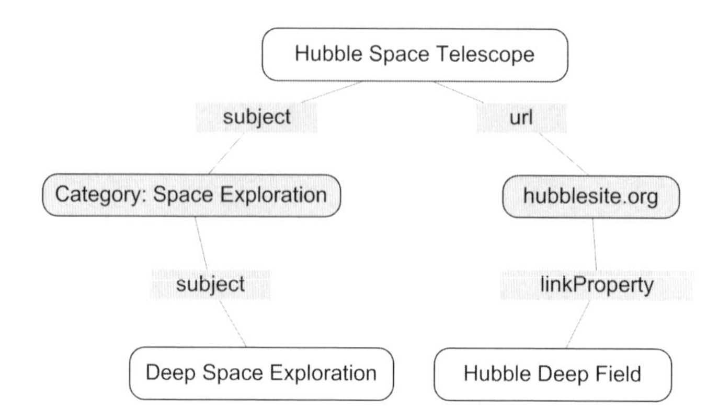

Fig. 3. Relation of relevant URIs for the examples tag set {hubble, deep, space, exploration, field}

Special Case – Simply one Ambiguous Tag in a Context: In case there is only one tag / keyword at a certain timestamp we have to create a context by comprising adjacent contexts.

The algorithm for detecting related entities generates a graph of entities. Related entities should be linked to each other and entities not applicable for the concerning context should stand alone. The set of entities derived from resolving ambiguous entities is disambiguated by detecting a path between the applicable URI from the set and other URIs from the context.

Ranking of URIs. The path-detecting algorithm produces n graphs of size $k = 1...m$. Graphs of size $k = 1$ will be deleted in case the concerning entity is retrieved from resolving an ambiguous URI, because there is no relation to any other element from the context. In case of an URI building an own graph of size $k = 1$ and not being retrieved from resolving the URI will be ranked very low in this context, because the URI is probably of low relevance for this context. The elements of graphs of size $k > 1$ will be ranked higher with increasing k. Within a graph the entities are ranked by the amount of in-going / out-going edges.

Additionally, the already disambiguated URIs are ranked by their co-occurrences. Wortschatz Leipzig provides amounts of co-occurrences for single words as well as composite word groups. The co-occurences of a term contain weightings that we are using to apply an additional score to our clusters of URIs. The ranking of URIs is then weighted regarding their co-occurrences and their relations in the LOD-Cloud.

Special Case – Name Components of Person Names: The group mapping algorithm maps single words and groups of 2 and 3 words to entities. Thus, in case of the context {peter, goddard} the algorithm will find mappings for "peter", "goddard", and "peter goddard". Certainly, the mapped URI for the whole name should be ranked higher than the URIs forname components. But, besides the whole name the URIs for the male name "Peter" and the family name "Goddard" should also be included in the set of assigned URIs to a video. In case a user searches for videos involving male persons in it or all persons with the family name "Goddard" these URIs need to be included rather than deleting them from the result set. This special case also affects the disambiguation process as the URI `<http://dbpedia.org/resource/Peter>` disambiguates to 31 further URIs, but in this case (and any other case the URI is referenced to a part of a person's name) only the URI `<http://dbpedia.org/resource/Peter_(first_name)>` is needed.

3.2 Indexation of Semantic Data

Entity-mapping and disambiguation are key problems in enabling a semantic search. Nevertheless, sophisticated indexing of the generated semantic data cannot be neglected to facilitate efficient search. This section shows that it is rather simple to extend a state-of-the art keyword-based search index with additional semantic information to enable an entity-based search.

The Yovisto Index. The Yovisto index is based on Solr[12], a Java open source enterprise search platform from the Apache Lucene[13] project. Major features are full-text search, hit highlighting, facetted search, and dynamic clustering. The indexing of documents requires the configuration of index fields, which make up the index schema. The schema specifies, which fields a document can contain and how those fields should be handled, when adding new documents to the index, or when querying those fields.

Corresponding to the structure of Yovisto's metadata, there are three different kinds of metadata:

(1) information referring to the entire resource such as the video title, subtitle, description, etc.,
(2) user generated tags assigned to a specific point in time of the video [5],
(3) time related text/keywords from automated analysis of the video (such as OCR on video frames).

In accordance with the type of metadata the indexing scheme has to be designed. For simple metadata (e.g. title) a single index field is sufficient to enable proper indexation. Time-dependent metadata carries the temporal information and is indexed either in one single field as (time, data)-tuple or in two dynamic fields with a mutual identifier, such as TIME_i, DATA_i. In case of user generated tags, the username has to be taken into account. Therefore, a (username, time, text)-triple is stored in one single field.

Fig. 4 shows the indexing structure of simple metadata, time-dependent user tags, and text/keywords in Yovisto. For every index document *single valued fields* are used for simple metadata (e.g. TITLE), *multi valued fields* are used for user generated tags (e.g. TAG), and *dynamic fields* are used to index more complex time-dependent data (e.g. SEG_*_TEXT). When the fields are indexed, analyzing

[12] `http://lucene.apache.org/solr/`
[13] `http://lucene.apache.org/`

```
(a)
TITLE:  "The Future of the Web"
DESCRIPTION: "We're approaching the end of 2010, and many people are
              wondering what the future will bring."
AUTHOR: "Cameron Chapman"

(b)
TAG: "[joerg, 31234] future"
TAG: "[joerg, 51231] internet"
TAG: "[nadine, 31215] future "

(c)
SEG_1_TEXT: "iPhone, MySpace, Facebook in 2025"
SEG_1_TIME: "121000"
SEG_2_TEXT: "No Print Magazines Have to Die"
SEG_2_TIME: "127000"
```

Fig. 4. The indexing scheme for simple metadata (a), time-dependent user tags (b), time-dependent text/keywords (c). The numbers represent the time point in milliseconds of the metadata within the video.

and tokenizing filters are applied to transform and normalize the field values. For example the *solr.WhitespaceTokenizer* creates tokens of characters separated by splitting on whitespace, and the *solr.PorterStemFilterFactory* applies the Porter stemming algorithm [10] on the tokens to enable to search regardless of inflection forms. The stemming filter is *not* applied to the tag fields, because an exact match is desired here.

Extending the Index with Semantic Data. As explained in section 3.1 for every metadata keyword a list of URIs with a corresponding rank is created. The rank indicates how confidently the URI was mapped to the keyword. To supplementarily index the corresponding URIs besides the original metadata, the idea is, to simply create additional fields. These additional fields comprise the URIs with corresponding rank and the corresponding position within the text data. In some cases it is even possible to extend the field instead of creating a new one. Note that adding new fields makes it very easy to adapt a regular search index scheme and extend it with the semantic information. Therefore, it is not necessary to modify the indexer implementation to enable a semantic supported search.

For every single valued field an additional URI field (TITLE_URI) is defined (cf. Fig. 5a). The user tag fields can be extended with the URIs. The text position can be omitted, because it would always be 0 (cf. Fig. 5b). The dynamic fields also require an additional field for the URIs, similar to the single value fields (cf. Fig. 5c).

To index the new URI fields properly, only the *solr.WhitespaceTokenizer* should be used. This is the reason, why new URI fields have to be created for origin fields with natural language values. In that case, usually other tokenizer

```
(a)
TITLE: "The Future of the Web"
TITLE_URI: "<http://dbpedia.org/resource/Future> 0.9432 1 <http://dbpedia
           .org/resource/Future_tense> 0.487 1 <http://dbpedia.org/resou
           rce/Web> 0.9342 4"

(b)
TAG: "[joerg, 31234] future <http://dbpedia.org/resource/Future> 0.9432
                     <http://dbpedia.org/resource/Future_tense> 0.487"

(c)
SEG_1_TEXT: "iPhone, MySpace, Facebook in 2025"
SEG_1_TEXT_URI: "<http://dbpedia.org/resource/IPhone> 1.0 0 <http://dbpe
          dia.org/resource/MySpace> 1.0 1 <http://dbpedia.org/resource/
          Facebook> 1.0 2 "
SEG_1_TIME: "121000"
```

Fig. 5. The indexing scheme extended with URIs, ranks, and text positions. Note, that the resource "Future_tense" has a lower rank than "Future" because the disambiguation algorithm considers "Future" more related to the resource.

and normalizers are used, e.g. word stemming. Since the user tag fields are only using the whitespace tokenizer, the URIs can be appended to the tag field values instead of creating a new field.

Querying the Semantic Data. Besides storing the data in an index structure the querying for an entity (URI) can be done by simply querying on the additional URI fields instead on the origin text fields. The result comprises all documents containing the requested URI. The document rank in the result is based on the Lucene document score. To map the user's query string to an entity an auto-suggestion widget can be used to support the user to disambiguate the query string by herself (c.f. Fig 6)[14]. The auto-suggestion list for URIs is

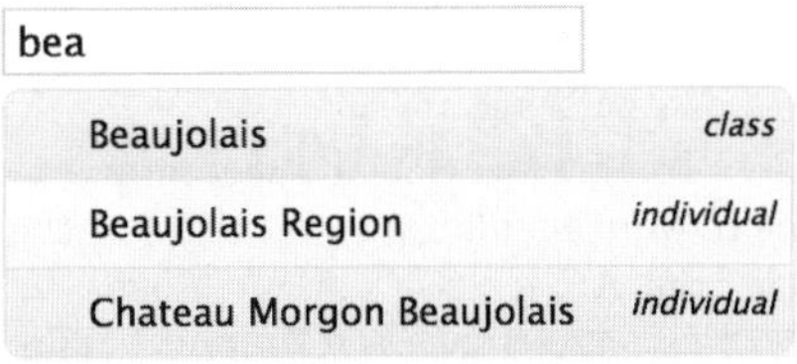

Fig. 6. Automated suggestion of entities while typing the user query. When the user selects an entity a query for its URI is issued, e.g. for "Beaujolais Region": `http://www.w3.org/TR/2003/PR-owl-guide-20031209/wine#BeaujolaisRegion` is issued. If the user does not select an entity, a regular keyword-based search is issued.

[14] The image originates from Jigs4OWL semantic mashup framework http://www.jigs4owl.com/

generated from the entities labels, names, and titles. If the user wants to search for more than one entity at one time, she can select more entities simply by continuing typing. This would provoke another suggestion list for the new query term.

4 Conclusion

In this paper, we showed how to transform a keyword-based video search engine to a semantic video search engine. We extended the existing Yovisto keyword/tag index by semantic information, which we determined by mapping the video content to semantic entities. Furthermore, we have explained how to store the newly-acquired semantic information in an efficient state-of-the-art search index without modifying its implementation or using special Semantic Web technologies. Finally, we provided the user a semantic search engine, that enables to search for LOD entities in the entire video repository of Yovisto without the typical shortcomings on keyword-based search.

The mapping process of keywords respectively user tags to DBpedia entities is a core problem in this research field and the challenging details of our approach are still under further development. To estimate how a user can be supported by providing a fully-fledged semantic search instead of a keyword-based search future work will in first place include a comparative evaluation of both search types. An evaluation of the entity mapping method with state-of-the-art benchmarks will be addressed in future work.

References

1. Berners-Lee, T.: Linked Data. World wide web design issues (July 2006), http://www.w3.org/DesignIssues/LinkedData.html
2. Bizer, C., Heath, T., Idehen, K., Berners-Lee, T.: Linked data on the web. In: Proc. of the 17th Int. Conf. on World Wide Web, pp. 1265–1266. ACM, New York (2008)
3. Carpineto, C., de Mori, R., Romano, G., Bigi, B.: An information-theoretic approach to automatic query expansion. ACM Trans. Inf. Syst. 19(1), 1–27 (2001)
4. Christel, M.G.: Supporting video library exploratory search: when storyboards are not enough. In: Proc. of the Int. Conf. on Content-based Image and Video Retrieval, pp. 447–456. ACM, New York (2008)
5. Sack, H., Waitelonis, J.: Integrating Social Tagging and Document Annotation for Content-Based Search in Multimedia Data. In: Proc. of the 1st Semantic Authoring and Annotation Workshop. Athens (GA), USA (2006)
6. Kleb, J., Abecker, A.: Entity reference resolution via spreading activation on rdf-graphs. In: Aroyo, L., Antoniou, G., Hyvönen, E., ten Teije, A., Stuckenschmidt, H., Cabral, L., Tudorache, T. (eds.) ESWC 2010. LNCS, vol. 6088, pp. 152–166. Springer, Heidelberg (2010)
7. Marchionini, G.: Exploratory search: from finding to understanding. Commun. ACM 49(4), 41–46 (2006)
8. Oren, E., Delbru, R., Catasta, M., Cyganiak, R., Stenzhorn, H., Tummarello, G.: Sindice. com: a document-oriented lookup index for open linked data. IJMSO 3(1), 37–52 (2008)

9. Pilz, A., Paaß, G.: Named Entity Resolution Using Automatically Extracted Semantic Information. In: Workshop on Knowledge Discovery, Data Mining, and Machine Learning, pp. 84–91 (2009)
10. Porter, M.F.: An algorithm for suffix stripping. Program 14(3), 130–137 (1980)
11. Prud'hommeaux, E., Seaborne, A.: SPARQL query language for RDF. W3C (January 2008)
12. Repp, S., Waitelonis, J., Sack, H., Meinel, C.: Segmentation and Annotation of Audiovisual Recordings based on Automated Speech Recognition. In: Proc. of 8th Int. Conf. on Intelligent Data Engineering and Automated Learning (2007)
13. Richert, M., Quasthoff, U., Hallensteinsdottir, E., Biemann, C.: Exploiting the Leipzig Corpora Collection. In: Proceedings of the IS-LTC 2006. Ljubliana, Slovenia (2006), http://wortschatz.uni-leipzig.de/
14. Schraefel, Wilson, M., Russell, A., Smith, D.A.: mSpace: improving information access to multimedia domains with multimodal exploratory search. Commun. ACM 49(4), 47–49 (2006)
15. Snoek, C., Sande, K.v.d., Rooij, O.D., Huurnink, B., Gemert, J.v., Uijlings, J., He, J., Li, X., Everts, I., Nedovic, V., Liempt, M.v., Balen, R.v., Yan, F., Tahir, M., Mikolajczyk, K., Kittler, J., Rijke, M.d., Geusebroek, J., Gevers, T., Worring, M., Smeulders, A., Koelma, D.: The MediaMill TRECVID 2008 semantic video search engine. National Institute of Standards and Technology, NIST (2009)
16. Stoermer, H., Rassadko, N.: Results of OKKAM Feature based Entity Matching Algorithm for Instance Matching Contest of OAEI 2009. In: OM (2009)
17. Tummarello, G., Delbru, R., Oren, E.: Sindice.com: Weaving the Open Linked Data. In: The Semantic Web, pp. 552–565 (2008)
18. Waitelonis, J., Sack, H., Kramer, Z., Hercher, J.: Semantically Enabled Exploratory Video Search. In: Proc. of Semantic Search Workshop at the 19th Int. World Wide Web Conference, Raleigh, NC, USA (2010)
19. Waitelonis, J., Sack, H.: Augmenting Video Search with Linked Open Data. In: Proc. of Int. Conf. on Semantic Systems 2009 (2009)
20. Waitelonis, J., Sack, H.: Towards Exploratory Video Search Using Linked Data. In: Proc. of the 11th IEEE Int. Symp. on Multimedia, pp. 540–545. IEEE Computer Society, Washington, DC (2009)

Towards the Automatic Generation of a Semantic Web Ontology for Musical Instruments

Sefki Kolozali, Mathieu Barthet, György Fazekas, and Mark Sandler

Centre for Digital Music, Queen Mary University of London, London, UK

Abstract. In this study we present a novel hybrid system by developing a formal method of automatic ontology generation for web-based audio signal processing applications. An ontology is seen as a knowledge management structure that represents domain knowledge in a machine interpretable format. It describes concepts and relationships within a particular domain, in our case, the domain of musical instruments. The different tasks of ontology engineering including manual annotation, hierarchical structuring and organisation of data can be laborious and challenging. For these reasons, we investigate how the process of creating ontologies can be made less dependent on human supervision by exploring concept analysis techniques in a Semantic Web environment. Only a few methods have been proposed for automatic ontology generation. These are mostly based on statistical methods (e.g., frequency of semantic tags) that generate the taxonomy structure of ontologies as in the studies from Bodner and Songs [1]. The algorithms that have been used for automatic ontology generation are Hierarchical Agglomerative Clustering (HAC), Bi-Section K-Means [2], and Formal Concept Analysis (FCM). Formal Concept Analysis is a well established technique for identifying groups of elements with common sets of properties. Formal Concept Analysis has been used in many software engineering topics such as the identication of ob jects in legacy code, or the identication and restructuring of schema in ob ject-oriented databases [5]. These works are important since ontologies provide the basis for information and database systems [6].

In this study, we present a novel hybrid ontology generation system for musical instruments. The music ontology is a Semantic Web ontology that describes music-related information (e.g., release, artist, performance), but does not provide models of musical instruments. Hence, there is a need to develop a separate instrument ontology to deepen how music knowledge is represented on the Se- mantic Web. Such complementary knowledge on musical instruments can be useful to develop music recognition and recommendation systems based on semantic reasoning. This work is a preliminary step which focuses on automatic instrument taxonomy generation in Ontology Web Language (OWL). The taxonomy of musical instruments given by Hornbostel and Sachs [3] was considered as the basis for our instrument terms and initial hierarchical structure. The hybrid system consists of three main units: *i*) musical instrument analysis, *ii*) Formal Concept Analysis, *iii*) lattice pruning and hierarchical form generation.

In the musical instrument analysis unit, the system analyses the relationships between 12 predened classes of musical instruments

T. Declerck et al. (Eds.): SAMT 2010, LNCS 6725, pp. 186–187, 2011.

(chordophones, aerophones, bowed, struck, reed pipe instruments, edge instruments, brass instruments, double reeds, single reeds, with valves, without valves, true utes) and 10 musical instrument individuals (bassoon, cello, clarinet, flute, oboe, piano, saxophone, trombone, tuba, violin) using a Multi-Layer Perceptron (MLP) classier. The underlying audio features used in the recognition and classication system are the Line Spectral Frequencies (LSF) which are derived from linear predictive analysis of the signal. The LSF characterise well the timbral differences between instruments, since they are related to the formant structure of the sounds spectral envelope which is an important aspect of the timbre. In this study, we used a database of isolated tones of various pitches and dynamics to train and test the system.

In the formal concept analysis unit, the binary relations obtained from the MLP outputs are used to identify the common attributes/individuals shared by different objects, and generate a concept lattice form. In the lattice pruning unit, the concepts are ordered in the hierarchical concept order and a well structured hierarchical form using reduced label ling technique is created. Finally, the reduced concept lattice labels and the class hierarchy of the instrument ontology is coded to the Ontology Web Language by using the OWL API java library [4]. The results have given a conclusive evidence in favor of the hierarchical similarity to the taxonomy of musical instruments used as a reference [3]. To the authors knowledge, this is the rst study to investigate automatic ontology generation system in the context of audio and music analysis. In further stages, the outcomes of this work may be used in an application to establish an automated mechanism to identify and classify music sources and annotate them with RDF meta-data. We will also incorporate a wider set of musical instruments and use more OWL language features (e.g., ob ject/data properties) which is important for Resource Description Framework (RDF) triples (subject S, predicate P, object O) and semantic reasoning. An instrument ontology is currently in development to be used as a ground truth for the further experiments.

References

1. Bodner, R., Song, F.: Knowledge-based approaches to query expansion in information retrieval. In: McCalla, G.I. (ed.) Canadian AI 1996. LNCS, vol. 1081. Springer, Heidelberg (1996)
2. Cimiano, P., Hotho, A., Staab, S.: Comparing conceptual, divise and agglomerative clustering for learning taxonomies from text. In: Proceedings of the European Conference on Artificial Intelligence (2004)
3. Doktorski, H.: (1996), `http://free-reed.net/description/taxonomy.html`
4. Horridge, M., Bechhofer, S.: University of Manchester and Clark & Parsia LLC and University of Ulm (2010), `http://owlapi.sourceforge.net/`
5. Snelting, G.: Concept lattices in software analysis. In: International Conference on Formal Concept Analysis, pp. 272–287 (2003)
6. Yahia, A., Lakhal, L., Bordat, J.P., Cicchetti, R.: An algorithmic method for building inheritance graphs in object database design. In: Thalheim, B. (ed.) ER 1996. LNCS, vol. 1157, pp. 422–437. Springer, Heidelberg (1996)

Video Analysis Tools for Annotating User-Generated Content from Social Events

Rodrigo Laiola Guimarães[1], Rene Kaiser[2], Albert Hofmann[2], Pablo Cesar[1], and Dick Bulterman[1]

[1] CWI: Centrum Wiskunde & Informatica
Science Park 123, 1098 XG Amsterdam, The Netherlands
{rlaiola,p.s.cesar,dick.bulterman}@cwi.nl
[2] Institute of Information and Communication Technologies, JOANNEUM RESEARCH,
Graz, Austria
firstname.lastname@joanneum.at

Abstract. In this demo we present how low-level metadata extraction tools have been applied in the context of a pan-European project called Together Anywhere, Together Anytime[1] (TA2). The TA2 project studies new forms of computer-mediated social communications between spatially and temporally distant people. In particular, we concentrate on automatic video analysis tools in an asynchronous community-based video sharing environment called MyVideos, in which users can experience and share personalized music concert videos within their social group.

In the MyVideos scenario, people attend to an event such as a school concert rehearsal and take videos, but not everybody films all the time. The purpose of each video is primarily personal: each family is interested in capturing their own child, but also enough context information from the concert to provide some background. Each family is interested in creating a video fragment for the personal family archives, or short clips that can be sent to family members who were not able to be at the show.

After the concert has taken place users can upload their video clips to the MyVideos platform. Before making such material available through a Web-based application, we use the Semantic Video Annotation Suite[2] (SVAS) and a video alignment tool [1] to automatically extract the metadata necessary to organize and tag the content subsequently. Besides generating potential key frames, the SVAS tool (see Fig. 1) detects severe unusual material that has the same appearance as shot boundaries – like very rapid camera movements, heavy unsteadiness or people crossing the picture close to the camera – and generates a set of annotations using the MPEG-7 standard[3], that then is stored in the MyVideos database. Other functionalities such as person and instrument recognition is currently under development – annotating user-generated content is challenging because the video encoding, the quality, and the lighting are not

[1] http://www.ta2-project.eu/
[2] http://www.joanneum.at/en/digital/avm/products-solutions-services/semantic-video-annotation.html
[3] http://mpeg.chiariglione.org/standards/mpeg-7/mpeg-7.htm

T. Declerck et al. (Eds.): SAMT 2010, LNCS 6725, pp. 188–189, 2011.

always optimal. Since users record, at their will, different parts of the concert, a key phase of the content preparation process is the alignment of all recorded video clips to a common timeline. For that matters, in addition to SVAS, a video alignment tool has been developed in the context of TA2 [1], and it can accurately align temporally user-generated video clips based on a high-quality audio stream recorded throughout the event. The results of such tools annotate the video clips and provide the means for easy search and navigation on the MyVideos front-end (see Fig. 1).

In our demo we present the MyVideos Web application (using an iPad) and the video analysis toolset (in the laptop), in their current, though not final, state of development. We give special attention to the automatic extraction processes of the SVAS tool, and we show how the metadata obtained allows for easy exploration of the concert media space (e.g., clip suggestion with relevant fragments within a clip for the user to watch or to share) and review of the video clips participants have recorded, along with their annotations.

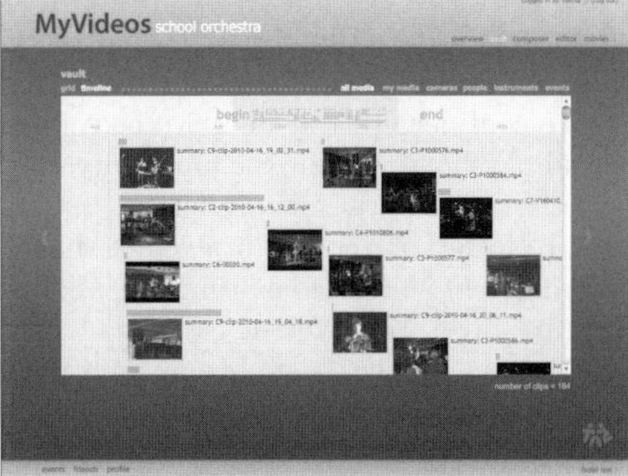

Fig. 1. Semantic Video Annotation Tool (on the left). MyVideos Web Application: Timeline View (on the right).

Acknowledgments. The research leading to these results has received funding from the European Community's Seventh Framework Programme (FP7/2007-2013) under grant agreement no. ICT-2007-214793. Special thanks to all the partners involved in the MyVideos demonstrator.

Reference

1. Korchagin, D., Garner, P.N., Dines, J.: Automatic Temporal Alignment of AV Data with Confidence Estimation. In: IEEE Proceedings of the International Conference on Acoustics, Speech and Signal Processing (2010)

Query by Few Video Examples Using Rough Set Theory and Partially Supervised Learning

Kimiaki Shirahama[1], Yuta Matsuoka[2], and Kuniaki Uehara[3]

[1] Graduate School of Economics, Kobe University
2-1 Rokkodai, Nada, Kobe, 657-8501, Japan
shirahama@econ.kobe-u.ac.jp
[2] Graduate School of Engineering, Kobe University
1-1 Rokkodai, Nada, Kobe, 657-8501, Japan
matuoka@ai.cs.scitec.kobe-u.ac.jp
[3] Graduate School of System Informatics, Kobe University
1-1 Rokkodai, Nada, Kobe, 657-8501, Japan
uehara@kobe-u.ac.jp

In this paper, we develop a video retrieval method based on *Query-By-Example* (QBE) approach, where a user represents a query by providing example shots. QBE then retrieves shots similar to example shots in terms of color, edge, motion, etc. We consider QBE as effective because the query is represented by features in example shots without the ambiguity of semantic contents. In addition, QBE can perform retrieval for any queries as long as the user can provide example shots. However, one crucial problem is that relevant shots to a query are taken in many different shooting environments. For example, as shown in Fig. 1, relevant shots are taken by different camera techniques and characterized by different backgrounds. Furthermore, people appear in different regions and face to different directions. So, even for the same query, relevant shots are characterized by significantly different features. Hence, our main research objective is to develop a QBE method which can retrieve a large variety of relevant shots using only a small number of example shots provided by the user.

Fig. 1 shows an overview of our QBE method, where the main algorithm is *Rough Set Theory* (RST) [1] which extracts multiple classification rules for discriminating between relevant and irrelevant shots to a query. RST requires two types of example shots, *positive examples* (p-examples), which are provided by a user to serve as representatives of relevant shots, and *negative examples* (n-examples), which are not provided by the user and serve as representatives of irrelevant shots. Thus, we formulate QBE as *partially supervised learning* [2] where classification rules are extracted only from p-examples, by collecting n-examples from shots other than p-examples. Our method iteratively enlarges n-examples by selecting shots which are more similar to already selected n-examples than p-examples. Such shots can be regarded as irrelevant with high reliabilities. In this way, a variety of n-examples can be accurately collected.

Then, we address the problem that the range of relevant shots which can be retrieved is inevitably limited using only a small number of p-examples. To overcome this, we use *bagging* where retrieval models are built on different subsets of randomly sampled p-examples and n-examples. In the case of a small number

T. Declerck et al. (Eds.): SAMT 2010, LNCS 6725, pp. 190–191, 2011.

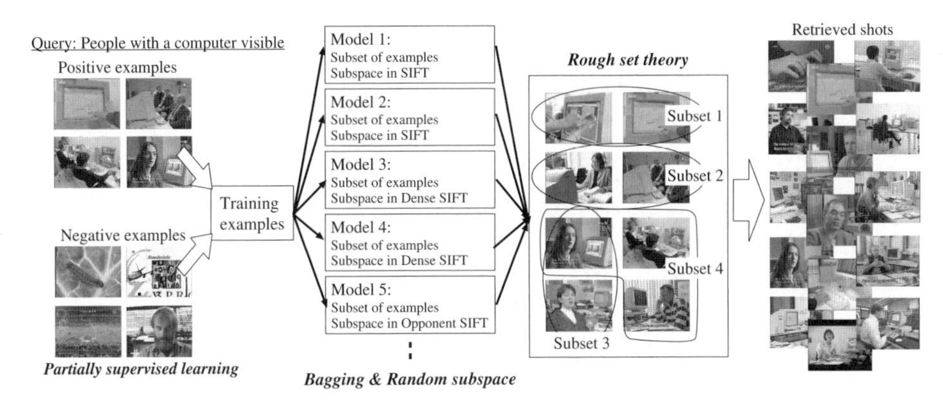

Fig. 1. An overview of our QBE method

of examples, retrieval results are significantly different depending on examples [3]. Thus, bagging is useful for extending the range of relevant shots that can be retrieved. Also, examples are represented by high-dimensional features, such as bag-of-visual-words representation. This causes that retrieval models tend to be overfit to feature dimensions, which are very specific to p-examples, but are ineffective for retrieving relevant shots. Hence, we use *random subspace method* where retrieval models are built on different subsets of randomly sampled feature dimensions, so that we can alleviate to build overfit retrieval models.

Finally, as shown in Fig. 1, RST is used to integrate various retrieval models built by bagging and random subspace method. Here, each classification rule is extracted as a combination of retrieval models, which can correctly discriminate a different subset of p-examples from all n-examples. By accumulating relevant shots retrieved by such rules, a large variety of relevant shots can be retrieved.

We test our QBE method on five queries. The result shows that when more than 30 p-examples are available, our method can successfully retrieves a large variety of relevant shots. Currently, to achieve the same level of retrieval in the case of less than 10 p-examples, we are incorporating into our QBE method an ontology as a knowledge base. In addition, for reducing the computation time, we are parallelizing the retrieval processe.

Acknowledgments. This research is supported in part by Strategic Information and Communications R&D Promotion Programme (SCOPE) by the Ministry of Internal Affairs and Communications, Japan.

References

1. Komorowski, J., Øhrn, A., Skowron, A.: The ROSETTA Rough Set Software System. In: Klösgen, W., Zytkow, J. (eds.) Handbook of Data Mining and Knowledge Discovery, ch. D.2.3. Oxford University Press, Oxford (2002)
2. Fung, G., Yu, J., Ku, H., Yu, P.: Text Classification without Negative Examples Revisit. IEEE Trans. Knowl. Data Eng. 18(1), 6–20 (2006)
3. Tao, D., Tang, X., Li, X., Wu, X.: Asymmetric Bagging and Random Subspace for Support Vector Machines-based Relevance Feedback in Image Retrieval. IEEE Trans. Pattern Anal. Mach. Intell. 28(7), 1088–1099 (2006)

A Reference Implementation of the *API for Media Resources**

Florian Stegmaier[1], Werner Bailer[2], Martin Höffernig[2], Tobias Bürger[3], Ludwig Bachmaier[1], Mario Döller[1], and Harald Kosch[1]

[1] Chair of Distributed Information Systems, University of Passau, Germany
[2] DIGITAL – Institute of Information and Communication Technologies, JOANNEUM RESEARCH, Austria
[3] Knowledge and Media Technologies Group, Salzburg Research, Austria

Keywords: W3C MAWG, interoperability, metadata annotation, standardization.

1 Introduction

Today many media sharing applications such as Flickr or YouTube exist on the Web that use diverse metadata formats to describe media resources. This leads to interoperability issues [1] in search, retrieval and annotation.

To address this problem, the W3C launched the *Media Annotation Working Group* [2], which aims to improve interoperability between multimedia metadata formats on the Web by providing an interlingua ontology and an API designed to facilitate cross-community data integration of information related to media resources on the Web. To do so, syntactic as well as semantic mappings between the so-called media ontology defined by the group and a large number of metadata formats have been identified in the group report *Ontology for Media Resources 1.0*.

2 Implementation of the *API for Media Resources*

This paper introduces a reference implementation[1] of the *API for Media Resources*, implemented as a Web service. The Web service is illustrated on the left hand side of Figure 1, highlighting its essential parts: the API defines interfaces which are exposed by the Web service and serve as an entry point for incoming requests (e.g., select a media resource). These could be sent by a non-UI agent (e.g., metadata crawler) or a user agent. Incoming messages rely on the media ontology and are translated into the underlying metadata formats by the use of the associated mapping rules. After the mapping, specific APIs as well as extractors perform the actual access to the data sources.

* The authors would like to thank all the participants of the *W3C Media Annotations Working Group* for their contributions. This work has been partially supported by the German Federal Ministry of Economics and Technology under the THESEUS Program and under the 7th Framework Programme of the European Union within the ICT project PrestoPRIME (FP7 231161).

[1] Demonstration online available at `http://mawg.joanneum.at/`

T. Declerck et al. (Eds.): SAMT 2010, LNCS 6725, pp. 192–193, 2011.

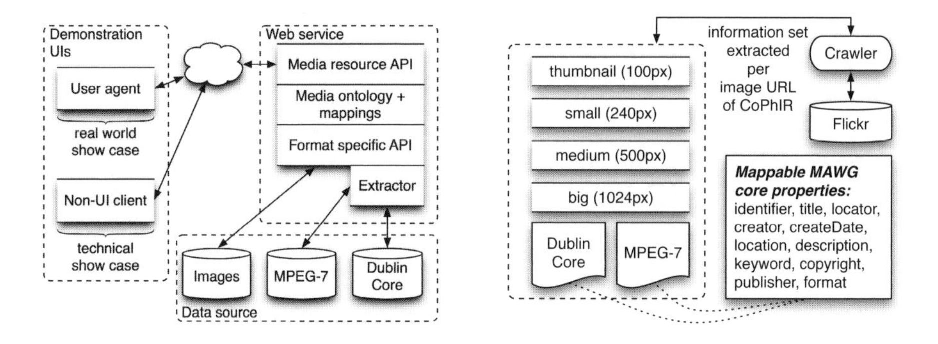

Fig. 1. Left: demonstration setup; right: data source, crawled using Flickr API and CoPhIR [3] as image URL archive

3 Demonstration Procedure

This demonstration shows how interoperability issues between metadata formats can be resolved by the use of the *API for Media Resources* (c.f. Figure 1 for the setup of the demonstration). The data used in the demonstration has been extracted from Flickr and consists of images annotated with MPEG-7 as well as Dublin Core. The right hand side of Figure 1 shows the configuration of the extracted data source. The demonstration illustrates the essential steps of the internal workflow: select a media resource, get available metadata formats, get (mapped) properties or show a diagnosis message at any time. In addition, the demonstration illustrates a real world image gallery use case. Here, the actual workflow (e.g., selection or metadata mapping) is hidden and only images and metadata information are presented to the user.

4 Conclusion and Future Work

This paper presents a demonstration of a first implementation of the *API for Media Resources*, using an image gallery application as a use case. It demonstrates the basic functionality of the API implemented as a Web service. In order to include additional metadata formats and mapping approaches as specified in the *Ontology for Media Resources 1.0* report of the group (cf. [2]), this work is continued as an Open Source project[2].

References

1. Smith, J.R.: The Search for Interoperability. IEEE MM 15, 84–87 (2008)
2. World Wide Web Consortium (W3C), Media Annotations Working Group. Video in the Web Activity, `http://www.w3.org/2008/WebVideo/Annotations/`
3. Bolettieri, P., Esuli, A., Falchi, F., Lucchese, C., Perego, R., Piccioli, T., Rabitti, F.: CoPhIR: a test collection for content-based image retrieval. CoRR, vol. abs/0905.4627v2 (2009)

[2] `http://mawg.sourceforge.net/`

Author Index